DISCARD

Cricket Radio

Cricket Radio

Tuning In the Night-Singing Insects

JOHN HIMMELMAN

THE BELKNAP PRESS OF HARVARD UNIVERSITY PRESS

Cambridge, Massachusetts

2011

Library of Congress Cataloging-in-Publication Data
Himmelman, John.
 Cricket radio : tuning in the night-singing insects / John Himmelman.
   p. cm.
 Includes bibliographical references and index.
 ISBN-13: 978-0-674-04690-0 (alk. paper)
 1. Katydids. 2. Crickets. 3. Katydids--Vocalization. 4. Crickets--Vocalization. I.
Title.
 QL508.T4H557 2011
 595.7'26--dc22
                                     2010035203

For the members of the Corps of Discovery

Contents

.

Preface

Sometimes the "how" comes before the "why." In the years I've spent getting to know the singing insects, I have managed to gather quite a bit of information about them. Writers often feel a responsibility to turn their obsessions into books. I turned mine into two.

In 2009 my book *Guide to Night-Singing Insects of the Northeast* was published. This is a field guide, illustrated with paintings by Michael DiGiorgio, meant to aid the curious naturalist in identifying these insects by sight and sound. It is the "how." What follows in the book you are holding is the "why." Why should we care to know more about the insects that are making these sounds? Why are they calling in the first place? The more answers I found to these questions (there's never just one answer), the more I discovered about the long cultural associations people have had with the singing Orthoptera.

Of course, I am not the first to undertake this. One of the inspirations for my pursuit in this area is the book *Crickets and Katydids, Concerts and Solos,* by Vincent G. Dethier. This engaging piece of literature takes you along on Dethier's journey as he gets to know these insects. I began his book a little curious about his subjects. By the time I flipped over that last page, they'd become a full-blown obsession.

The title *Cricket Radio: Tuning In the Night-Singing Insects* is not my own creation. It came from Ann Downer, a former editor for Harvard University Press. It was inspired by the chapter about a young boy who would be lulled to sleep at night by the songs of crickets and katydids. It was not enough to hear them through an open window. He and his father set out to capture them and bring them into the boy's room, where the calls would take on the role often relegated to a radio or television. Something about that spoke, or sang, to Ann, and she encouraged me to pursue it further. There was no doubt that boy was on to something.

While I have included photos representing most of the Ensifera (katydid and cricket) families and many of the species in North America, *Cricket Radio* is not a field guide. It is my hope that having read this book, you will be inspired to go out and learn more about these insects. At the end of this Preface I've listed the other guides that are available, besides mine, and I recommend that you seek them out.

Some of you, however, will be perfectly happy to just enjoy the sounds you have been reawakened to, without needing to know the name of every single singer. Consider yourselves among the fortunate who can pull from the night a gift for the ears, freely given.

Digital technology has allowed me to share with you some of the sounds *I* hear. The songs in the Online Audio Resource (which can be accessed at www.cricketradiobroadcast.com.)are meant to engage a sense that can be given only limited expression in the written word—hearing. We all have our way of hearing things, and describing what we heard. You can listen to, in their natural settings, many of the species I write about in this book. I guarantee that you will have heard a number of these sounds before. My hope is that you will find it nearly impossible to ignore those sounds when you hear them "in person" on those warm summer nights.

Recommended Field Guides

Guide to Night-Singing Insects of the Northeast (CD with calls included), by John Himmelman and Michael DiGiorgio (Stackpole Books)

Field Guide to Grasshoppers, Katydids, and Crickets of the United States, by John Capinera, Ralph D. Scott, and Thomas J. Walker (Cornell University Press)

The Songs of Insects (CD with calls included), by Lang Elliott and Wil Hershberger (Houghton Mifflin)

Field Guide to Insects of North America, by Eric R. Eaton and Kenn Kaufman (Houghton Mifflin)

"Singing Insects of North America"—http://entnemdept.ufl.edu/walker/buzz/

Cricket Radio

one

Music Beckons

> Most haunting of all is the one I call the fairy bell-ringer. I have never found him. I'm not sure I want to. His voice—and surely he himself—is so ethereal, so delicate, so otherworldly, that he should remain invisible, as he has through the nights I have searched for him.

The actress portraying ecology movement pioneer Rachel Carson spoke these words in the 2009 film *A Sense of Wonder.* They were taken from an essay she had written in *Woman's Home Companion* in 1956, called "Help Your Child to Wonder." The passage recounted time spent searching, with her grandnephew Roger, for the sources of the "insect orchestra" that swelled and throbbed outside her Maine cottage from midsummer until winter.

Carson writes, "The game is to listen. Not so much to the full orchestra, as to the separate instruments, and to try to locate the players."

She describes her fairy bell-ringer's call as clear and silvery, and faint. It is "so-barely-to-be-heard, that you hold your breath as you bend closer to the green glades from which the fairy chiming comes."

While I'm sure the scientist in Carson really *did* want to find that little "fairy," I can understand why another part of her didn't mind leaving it a mystery. There is something to be said for enjoying something for enjoyment's sake. For the perpetually curious, it can be a challenge to override that part of you that needs to know more about those sources of wonder.

Sometimes, as with the crickets she was unable to find, that choice is made easier for you.

I've no doubt she knew those crickets were not calling for her pleasure. They were hard at work at the business of holding a place for their kind on this planet. There could be no more urgent and consequential task to be undertaken by those insects, or for any creature. There was no joy in their song. There was no celebration; nor was there sadness or sorrow. Those "players" were throwing everything they had into propagating their species, and not by choice. They are hardwired to do so. Yes, Carson knew that, but she probably also knew that there is a kind of beauty in those cold, hard facts. It is like the mathematician waxing poetic at the austere elegance of the Pythagorean theorem. Physicists find beauty in the makeup of mass, matter, and motion. I suppose the common theme is balance and harmony—things working as they should. A cricket rubbing its wings is carrying out its purpose, as it should. There is something added to that, though. Sound has a way of stimulating our brains. It enters through our ears and resonates within the auditory cortex. The hippocampus, responsible for long-term memory, is located just below that auditory cortex. It integrates with this region, adding connections to our past, along with the associated emotions.

That beauty we derive from the songs of crickets, birds, and whales, the enjoyment of listening to waves crashing on the shore, or wind through the leaves, comes from our own perception, interpretation, and triggered memory connections. We are hardwired, too. It is in the design of the human spirit to be stimulated by things not undertaken for our own edification.

I remember when I first came to the realization that some of the most uplifting and ethereal moments of nature occur while watching her struggle to survive. I was hiking down a trail in a white pine forest on a cold January morning. The songs of Black-capped Chickadees and Tufted Titmice whistled, chipped, and buzzed around me. It was such a brisk and wintery sound, and their song lifted my spirits. Why? Because they sounded . . . *happy.* I watched them fly from branch to branch, in constant communication with one another. It then occurred to me, "It's *really* friggin' cold out here!" These birds need to stay strong to keep their body temperature up. They need to eat. They need to find food! If they did not find food, they would grow weak and die. The calls they produced aided in that effort. It kept them together. Flocking increases the chances that some in their group will find food for all.

People whistle when they are happy. Chickadees and titmice, while struggling to maintain their energy level on a freezing winter morning, whistle to survive. In that moment, those cute little, carefree birds transformed into hardy, focused animals with little time for the flights of fancy their human observer enjoyed.

My view of them took a radical turn. But surprisingly, it didn't matter. My newfound appreciation for the struggle of those little birds had no effect on my appreciation of their song. To my ear, it was no less ethereal. In fact, I liked it even more. I'm sure Rachel Carson shared a similar epiphany in her life. I would think she almost would have had to. Any scientist who can devote so much time to a subject, hers being the outdoors, has to be tickled by the magic.

I will admit that in listening to the actress describe the night hunt in that film, part of me was basking in the glow of her lyrical words, while another part my mind was racing to figure out what crickets she may have been listening to. My hippocampus was abuzz with memories of bell-like calls and the bugs that produced them. I don't know if they make it that far north, but I kept narrowing the callers down to Say's Trigs *(Anaxipha exigua)* or Tinkling Ground Crickets *(Allonemobius*

tinnulus). These tiny crickets do produce a tinkling call, pleasant enough to the ear to drive one to seek them out.

As much as I'd like to believe otherwise, the Ensifera (katydids and crickets) don't enjoy their songs in the way we do. Nor
do the birds or frogs. If that's the case, then why sing? There
is a lot of energy expended in song production. In one night,
a singing Spring Peeper will have spent the same comparative
amount of energy as a person running the New York City
marathon. Energy must be replenished with food, the search
for which burns up even more precious fuel and puts them in
the path of predators. I have a vegetable garden right outside
the window of my studio. Spring, and then Fall, Field Crickets
(Gryllus veletis and *pennsylvanicus)* call from early afternoon
into the night. Imagine yourself doing *anything* for that long.
Granted, when it comes to the katydids and crickets, their food
is usually just beneath their feet. While many of them supplement their meals with tiny arthropods, they do not chase them
down like lions hunting a zebra. Those creatures, too, are in
their vicinity. However, what if they didn't have to replenish
their energy from hours-long stints of singing? Wouldn't that
be an advantage worth passing on to the gene pool? Especially
if, as I surmised earlier, being insects, they don't have the
appreciation for the sounds they create?

Insect song has been around a long time on this earth. Fossil
evidence exists of sound-producing, sound-sensing Ensifera
forebears as far back as the late Permian, about 250 million
years ago. The stridulary (sound-producing) apparatus and ear-
on-elbow design has changed little over eons. Because of this,
we can imagine the sounds forming the backdrop of a hunting
Allosaurus, or a grazing Brachiosaurus. It is the sound of two
hardened wings being rubbed together, multiplied many times
over.

It is hypothesized that insect song began as visual signaling, which in turn could have arisen from the male's act of lifting his outer wings (tegmina) to allow the female to mount from behind. Lifting the wings from the body was likely an early display employed by the ancestors of modern Ensifera. Those wings would be flashed or vibrated to call more attention to them. Perhaps, in addition to advertising for a female, the flash signals were used to intimidate other males. It could be a size advertisement, meant to urge the interloper to back off. Visual displays are preferable over an actual fight, where both loser and winner can suffer severe damages. We see such visual displays today in birds, reptiles, and many insects. Over time, certain mutations added tiny bumps to those outer wings. The act of closing those wings made a little noise when those bumps rubbed against the other wing. The early noise-producing insects were likely to be relatively quiet, suggesting this new adaptation was only effective within a close range. This should be safe to surmise, because the hearing apparatus was still developing. What would be the benefits of passing on the traits for producing sound, if there were no way to hear the product?

The songs of those earliest sound-producing insects were fairly modest, likely composed of simple clicks and scrapes. As the wings developed more teeth for the files, and more efficient areas for sound amplification, those songs grew more complex. Female insects that could better locate calling males from a distance passed on their improved hearing traits. Males also benefited from being able to hear their competition. It pushed them to prove their worth with the most strident song they could muster.

In Chapter 3, I discuss how the Ensifera calls are produced by dragging the file of one wing across the scraper of another—much like a fingernail over the teeth of a comb. This is done rapidly, and the sounds we hear are actually a series of pulses. The insect's brain sends a signal to a pattern-creating ganglion in the thorax. That signal can be triggered by factors such as sperm availability, temperature and other climate factors, and light intensity. The latter plays a role in designating insects as day, dusk, or night callers. That ganglion of nerves switches on the muscles to set the wings in motion to produce that sequence of pulses. In species that trill, those pulses are very close together. In species that merely "tick," they are farther apart. The rhythm conformity of those pulses varies from species to species. Some, like the tree crickets, adhere to a steady rhythm. Others, like the Fork-tailed Bush Katydid *(Scudderia furcata),* give a less unified pattern of ticks.

We can distinguish several categories of Ensifera song, though perhaps referring to their sounds as "signals" would more accurately describe their function. "Song" can suggest a purpose that's different from what these insects are doing. For crickets and katydids, their sounds are auditory signals, designed to convey an important message. What follows are four of the main categories of those signals:

1. Calling song. This is the loudest, most sustained, rhythmic, and intense call of the fertile, adult male. It is meant to cast a wide net for females, and it is the sound with which we are most familiar. The males give this call from their territory, which is often defended. The call, in addition to luring females, helps establish that territory. The calling song of one male usually stimulates other males to call. A lone male in captivity will usually sing less than one surrounded by others.

   Where the insect calls from is very important, and many species maintain a certain spatial relationship with other calling males. If they get too close, one insect's call can interfere with another's. If they are too far from other calling males, they may miss out on the benefit of more females being drawn to an area.

In males that are solitary, as opposed to those that chorus with other males, the intensity of their song is the greatest factor in attracting females. Those that call the longest and loudest get more attention.

2. Response song. This sound is less of a song, and more of a brief call. It is given by some female katydids in response to a calling male. It is generally shorter and more subdued than the male call.

3. Courtship song. Male crickets use this while in close proximity to a potential mate. It is a softer and more intimate call, and might be used while the male is attempting to entice the female to mount him. It is believed that calling songs evolved from courtship songs. Some of the Phaneropterinae katydids have a particular call designed to elicit the response of a nearby female. Once the male hears the nearby female, he may continue to call, but will ease up its intensity. Many katydids, however, employ only the calling song.

4. Fighting/defensive song. In both crickets and katydids, there can be an instance when one male gets too close to another. The intruder is told to back off by a brief, sharp, nonmusical click, a series of them, or many of a variety of pointedly aggressive calls. Some katydids give a short click when captured, to startle a predator into dropping it.

In the Ensifera world, it's the males who call the females hither. This is the same in the amphibian and avian world. In the moth world, it's the other way around. The females set adrift a lure of pheromones. The males take wing in an attempt to pick up their scent with the aid of their feathery antennae. If he's able to trace it back to the source, the male then releases some pheromones of his own to reassure her they are of the same species. Once those proper introductions are out of the way, mating takes place. This is obviously a larger expenditure of energy for the male, who is taking a shot in the dark at picking up the scent. It is also puts it at increased risk of getting caught and eaten. Greater mobility puts them in the path of more predators, such as bats and nocturnal avian insectivores. I sometimes run a UV light in my yard to sample the moth diversity in my area. While I do come across females, males make up the greater percentage of species I attract. None of these were goaded to leave their safe hiding place for the glowing light. They were already on the move and passing within the range of that light. The light that picked them off could have just as easily been a predator.

In the world of flies, the midges, blackfly, mayfly, and mosquito males are again the risk takers. They gather in larger mating swarms, forming huge "Eat Here!" billboards for dragonflies and birds.

Of course, we all know what the male praying mantis sacrifices in the name of continuing his lineage—his head. While this is not always the case, female mantids have long been known to devour the head of their mate during copulation.

Male frogs, too, run a risk in taking the more active lead in courtship, particularly in the tropics where certain bats and opossums have acquired a taste for them. Those mammals use auditory clues to locate their amphibian prey. Some larger frogs will home in on singing males of smaller species, upon which they feed. The next time you hear a chorus of Spring Peepers, try to approach them. You will notice that the second you are seen or heard, they stop calling. The silence is a survival response. It is employed by animals that have evolved alongside predators who hear their calls as dinner bells.

There can be no doubt that the males in these scenarios, in addition to expending the greater energy in mate seeking, are taking a larger risk. It is no easier for the katydid. The males do the calling, and *when* they call may have something to do with when the females are receptive. They could sing all they want in the darkest, safest, most out-of-the-way, predator-free area they can find, but if there are no females responding, what's the point?

Sure, a lot of them, especially the katydids, look like a leaf. However, looking like a leaf does an insect little good if that leaf is moving in a very "un-leaflike" manner, *and* making very un-leaflike noises. A true or angle-winged katydid, calling away on top of a leaf, as they do, may be bringing females closer, but it is also bringing unwanted attention to itself. Bats are known to home in on the calling insects. If the katydid hears the bat's sonar in time, it will stop calling. Katydids have been on the evolutionary journey with the bats long enough to recognize their threat. Despite this, many katydids become bat food.

In the meadows, a stridulating male conehead (a type of katydid) or cricket risks a similar threat of being located. There are a number of insect hunters on the wing, including one group, the ormiine tachnid flies, that are listening for them! They actually possess tympanal hearing organs for just this purpose. This adaptation was discovered by Dr. William Cade, of Alberta, Canada, in 1975. The fly seeks out the host by listening for its call, and then sprays it with its first instar larvae, or *planidia*. The larvae burrow into the host, and feed within, eventually killing it. Research has shown that the presence of these sound-targeting flies has the ability to alter the calling patterns of certain crickets. The crickets that call with shorter, less frequent bursts have a better chance of eluding the flies. Naturally, the key to this working involves the females' cooperation. They have to respond to the abbreviated call, which, to the relief of the males, they do.

Thomas Walker of the University of Florida noticed domestic cats going after singing katydids and crickets. He conducted an experiment using the recorded calls of a katydid the subject cat would be unfamiliar with.[1] The cat went directly toward the speaker. Walker stopped the call, and the cat sat there, staring up at where the sound came from. During one session, the cat started toward the katydid call in the speaker, but when it was about five feet away it went after a moth crossing its path. If the katydids can somehow figure out how to attract moths with their call, they may buy some safety—at least from the 70 million feral cats in the United States.

The worst way to go, if you are a katydid, is death by digger wasp! Frequent victims of theirs are the meadow katydids. Again, a calling, moving katydid is a more noticeable target than a relatively still, skulking female. These males are trying to be found, so they have to take certain risks. Although the digger wasps do not listen for their prey, they search for them through the low vegetation. The wing movements of a calling male could make it stand out more than an insect sitting perfectly still. The wasp descends upon the victim, stinging it until it is paralyzed. The katydid is then carried aloft to the entrance of the wasp's underground burrow. The antennae are chewed off, to make it easier to drag into the hole, and the katydid is hauled down to its awaiting hell. The wasp then lays her eggs in the chamber and moves on. When the eggs hatch, the larvae feed on the paralyzed, but still living, host.

The same method of reproduction is practiced by the grass-carrier wasps, which are in the same Sphecidae family as the digger wasps. These are the ones you see flying with a piece of grass in tow. They line their tunnel (often using the tracking in your aluminum screens) with the grass, and provide for their young in the same manner as the digger wasps. The difference lies in the choice of hosts, which are often tree crickets or Drumming Katydids *(Meconema thalassinum)*. In my area, the Narrow-winged Tree Crickets *(Oecanthus niveus)* seem to be the cricket of choice. Because these crickets call primarily at night (the Drumming Katydids, not at all), I have not seen an abundance of male over female hosts in nests I've checked, and I've checked quite a few. I'd say this negative evidence might add a tendril of support to the dangers inherent with calling. If the male is silent in the presence of the predators, its chances of becoming captured become no more than an equally silent female's.

This method of Hymenoptera reproduction, by the way, has caused much discussion over the centuries on the existence of God, the question being: How could a benevolent and loving God design, or even allow, such a horrifying interchange among his beloved creatures? Even Darwin puzzled over this. In 1860 he wrote a letter to his friend Asa Gray, a botanist and devout Christian, in which he said:

> There seems to me too much misery in the world. I cannot persuade myself that a beneficent and omnipotent God would have designedly created the Ichneumonidae with the express intention of their feeding within the living bodies of Caterpillars, or that a cat should play with mice.[2]

Granted, the digger and grass wasps are in a different family from the Ichneumonidae. From a hosting Orthopteran's point

Drumming Katydid (*Meconema thalissinum*), female.

of view, however, it's just semantics. Ichneumonids, instead of having the larvae chew its way *into* the body, save it the trouble and inject the eggs directly into its host, which, when hatched, chew their way *out.*

If ever there was a reason *not* to sing . . .

The bottom line is that enough individuals elude predation to make the act of calling worth it. Because males are relegated to the station of risk takers in courtship, it is their sex that has evolved the better apparatus for signaling. Remember, though, many female katydids also stridulate. This act seems to be for the purpose of allowing an already nearby male to know it is there. It may even be a way of informing him that they are of the same species, much like a male moth's pheromone assurance. John D. Spooner has observed that although the females are drawn to the call of the male, the males of some species travel to the female in response to her call.[3] They play a form of "Marco Polo," he issuing his quieter call, she responding with ticking, until they meet up. This could reduce the calling time of the male, which, in turn, reduces the amount of time he is at risk from predators and parasitoids. So, it appears that the females in some species do care . . .

Assigning the males to the role of higher risk taking illustrates the greater importance evolution places on the female when it comes to propagating the species. The females bear the eggs. If you lose a female, you lose hundreds of eggs. If you lose a male, there will be another right behind it with its tarsus raised, saying, "Put me in, coach!"

That is not to say that, in addition to meeting a female halfway, males don't exercise other precautions in their courtship process. The first response when faced with danger is to simply stop calling. Most insects, or animals, for that matter, go silent at the approach of a potential predator. They also stop moving. Katydids, if still threatened, drop to a lower leaf or into the tangle of weeds. Some will fly. Crickets go under things. Coneheads scuttle to the base of the stalk and do their "piece of grass" imitation. The Ensifera ear serves several purposes. One is to hear each other call. Males hear other males and gauge their proximity to one another, allowing them to set up their own calling station. Katydid males also hear responding females. Females hear males call, which allows them to locate them. However, there are some Ensifera that do not call, and yet have ears. This suggests that the hearing apparatus was retained to clue them in on the presence of predators. The ability to hear danger must be a positive trait in an insect with few, if any, ways of fighting off enemies.

Some of the Ensifera attempt to confound predators by limiting the volume and length of time they call. It becomes a balance of calling long and loud enough to be heard, and yet briefly and quietly enough to escape predator detection. This is behavior molded by the predators.

Calls of insects have evolved to find such balance, and continue to change as new challenges arrive. First, the call must match the habitat. Certain sounds work better in certain settings. Some calls will need to bounce around among the stems and blades of grasses and sedges. Some will be set aloft from an open perch, while others need to resonate from an underground burrow.

No matter where they are, the Ensifera will fall prey to creatures tracking them by sound and/or movement, but it doesn't mean they haven't been working on the problem. I was thinking about some of the crickets and katydids that call from somewhat exposed perches—some examples being the true and bush katydids, and the angle-wings, that stridulate from the tops of leaves. It occurred to me that they are primarily nocturnal. While there are bats, and sometimes owls, to contend with, there are fewer predators on the wing than there are during the day. I would imagine that if these insects were equally active when most of the birds and wasps were up and about, there'd be way more songs cut short.

Then there are those, like the meadow katydids, that call during the day. They are active when most of the birds are active. However, they are mixed in deep among the tangles of monocots, clinging closely to them, keeping a low profile. Their biggest problems are other insects sharing that intimate habitat. Many ground and field crickets sing during the day, too, but look at where they reside—under things. Many will come out to the surface to call, but they are just a quick hop from cover.

As far as niches go, nature abhors a vacuum, and there will be an insect to fill nearly any habitat. If that insect happens to produce sounds necessary to the survival of its kind, then it stands to reason that those sounds will have evolved to suit that niche.

In addition to working within the habitat, the calls have to say something about the males making them. To start with, that call has to be specific for a species. This means there needs to be a degree of consistency with regard to pitch, rhythm, volume, frequency, and when it is given. Fortunately for us, differences in most of those calls are recognizable to the human ear. It is possible to walk through any habitat and list the species based upon hearing their unique stridulation. A number of the katydids produce ultrasonic signals that the unaided human ear will never hear. Some add them to their audible repertoire, while some use them as the primary call.

The crickets and katydids do not learn these songs from elder Ensifera. They are assigned them following the last molt. Experiments in the laboratory have shown that even an Ensifera deafened before its first song, can still sing the correct tune for its species. Another experiment was undertaken where the late instar nymphs were continually fed the sounds of another species. When they emerged as adults, their song was no different from any other of their kind.

The sounds they produce, recognizing aberrations, of course, are limited to the morphological consistencies of a species. One insect can have several calls, but so too will all the other like-sexed members of that species. The only variations of those sounds result from temperature changes, aging and wear of the stridulary organ, and the minor size differences within a species. The health of the caller also plays a role, and it is something a female attempts to measure when sizing up a potential mate.

The mating call of the male becomes a chicken-or-egg thing. Do the males shape the female's choice in song, or, in selecting certain singers, is the female shaping the male song? Research has shown that female Ensifera tend to gravitate toward larger males. Perhaps the greater size suggests health, or the ability to hold a territorial claim. It can also be an indicator of a greater nuptial offering (as discussed in Chapter 3). A larger male will produce a call commensurate with its size. It is something that cannot be aped by a smaller male bearing smaller stridulary apparatus.

Males also use their calls, somewhat indirectly, to control other males. One male calling will spur another to do the same. If the first male stops, it can cause the other to stop as well. If it speeds up, or becomes louder, the neighboring males may endeavor to match him. The calls of males also act as a kind of auditory force field to keep them far enough apart from one another. Although they remain in "earshot," they are usually not in visual contact. This reduces time (and danger) spent physically engaging competitors and may allow females to approach an individual male without being overwhelmed by a horde of suitors.

Finally, the calls have to place the caller in as little danger as possible. As I mentioned earlier, some species modulate their calls to escape the attention of predators. Some, however, go another direction and let loose with all they've got. These could be individuals hidden well enough to avoid detection.

They could also be part of a chorus.

Many members of the animal kingdom join in chorus, usually for the purpose of attracting the opposite sex. We hear the communal gathering of Spring Peepers, peeping in unison in the spring, and know those are the males calling in the females. This is followed later in the season by the synchronized trilling of Gray Treefrogs, linked in song for the same purpose. What's a dog-days afternoon without the joined drone of cicadas? Unless you happen to be standing beneath one of them, it is very difficult to distinguish individual callers from the overall cacophony.

Chorusing can take place without sound, too. If you find yourself along the coastal plains in Georgia between May 10 and May 28, you can watch male fireflies *(Phoruis frontalis)* blinking in unison. As long as you're on the coast, hop a ship to Portugal, specifically to the Ria Formosa National Park. You will feel most welcome there, as the male European Fiddler Crabs wave in greeting. They are not waving at you, though, but at the female crabs. Groups of male crabs gather and brandish their large claws in unison, each hoping the females will notice his is bigger than that of the crab next to him.

I am not being wry when I include humans among nuptial choristers. Think of the answer given by the stereotypical rocker when asked why he started a band—"For the babes." He joined his sound with other musicians hoping to draw a female fan base. It is his hope that one of those females will select him from the other members of his band. Often, as you will see with the katydids, it is the lead singer.

Music beckons.

Many of the cricket and katydid males join their wings in song with neighboring males of the same species. This chorusing can be synchronized or alternated. One of the best-known synchronized Ensifera choruses is provided by the Snowy Tree Cricket *(Oecanthus fultoni)*. These delicate ivory-colored singers take great pains to match the pulsing of their stridulation with the crickets around them. The act creates what sounds like one humongous throbbing Snowy Tree Cricket. Columbian Trigs *(Cyrtoxipha columbiana)* sing in sweet unison from the upper canopies of southern trees. In the marshes, the shushing calls of Long-beaked Coneheads *(Neoconocephalus exiliscanorus)* join in rhythm. Bentley Fulton shared an observation about Long-beaked Coneheads by his colleagues James Rehn and Morgan Hebard: "[They] described the song as a vibrant rattling note rising and falling in intensity often ceasing as if in exhaustion."[4] He went on to explain that this variation and intensity were

Long-beaked (aka Slightly Musical) Conehead *(Neoconocephalus exiliscanorus)*.

not features of the individual song, but were due to the changes in the number of singers participating.

Common True Katydids *(Pterophylla camellifolia)* take turns, alternating their calls with a nearby katydid. What happens is this: One male leads off the call. A neighboring male joins him, which causes that leading male to slow down his rhythm to about half as fast. The neighboring male matches his speed, but fills in his silent gaps with *his* croaking beat. To our ear, this creates a rhythm similar to that of the original call of that lone, lead katydid. If you have a large colony, each matching a neighbor, who's matching a neighbor, who's matching a neighbor . . . their calls meld together, beating like the giant heart of the dark forest.

Richard Alexander wrote a curious account of a true katydid chorusing with a typewriter.[5] A male in captivity was heard calling during the day (they are primarily night singers) when someone was typing in another room. To see if the typewriter was causing this, he tapped the keys in the rhythm of the katydid. The katydid slowed down its phrase and immediately

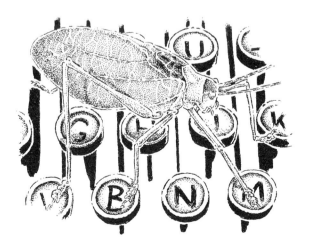

began to alternate its call with the typewriter! Alexander played around with this for a few hours and learned that the katydid would always respond to the typewriter, which, in effect, became the lead singer. When he typed slowly, the katydid responded in kind. When he typed rapidly, the katydid waited until the typing stopped. Then it took its turn, stridulating in a rapid rhythm.

Alexander concluded that a true katydid sings sooner after being stimulated by a phrase by another katydid, than he does after a phrase he produced himself. This behavior is key to a consistent alternation. It's why they're so good at it.

Some of the Orchelimum (larger meadow katydids) alternate their calls as well (and some synchronize). So, too, do the Rattler Round-winged *(Amblycorypha rotundifolia)* and Carinate *(A. carinata)* Katydids. Last summer I stood in a field, surrounded by calling Carinate Katydids. Each individual was scratching out his high, harsh, slow, and steady series of "zits"— but alternating with the others. The effect was like sitting in a surround-sound theater. The calls shot from front to back, from side to side, each following the cue of the previous caller. The combined effect from a distance is that of a long, uninterrupted call that an individual insect would be hard pressed to maintain, but produced seemingly without effort by the chorus.

There are many advantages to chorusing in the Ensifera world. For the males, it allows the opportunity to create a greater curtain of sound. A single call could get lost among the mélange of insect calls; the call of many creates a larger target for the male-hunting female. It also allows for maintaining the rhythm of a particular call. A colony of stubborn individualist singers could sound to a female like an elementary school band warming up their instruments. Singing together, however, they emphasize the elements of their song females find so fetching.

From the female's point of view, she is heading to a place where she can do some comparison shopping for males. Instead of risking the trip for an individual, who may not pan out, she can sample the wares from many individuals within a relatively small area. It is not unusual for a female to climb onto the back of a prospective mate, only to climb back down again after deciding he's not the *one*. Many species opt for the stronger singer of the group, or the individual that the others of the chorus are following—that lead singer.

In addition to increasing the overall decibels of their lure, the males can have the added benefit of increased duration and intensity of the call. One male can take a break while others continue to cast the net, uninterrupted. When I was a kid, I'd do that in chorus when I'd forgotten the words (keeping my mouth moving, of course). No one was ever the wiser.

Another advantage offered by chorusing is gained by the individual within that chorus. There is safety in numbers. I visit elementary schools with a workshop on writing about nature. A naturalist friend of mine, Frank Gallo, shared with me a lesson he does with kids. I tweaked it a bit, and now offer it to the students I visit. It goes like this:

One child plays the katydid (it's usually a girl named Katie, as there's almost always a Katie in an elementary school class-room). Another child plays the hungry bat. The "bat" child is blindfolded and is given the task of finding the "katydid" child, who is told to clap every few seconds. The bat has little trouble locating her by sound. Then eight kids join in a circle with the katydid caller. They will form the *chorus* of katydids. The blindfolded bat is in the middle of the circle. I give each "katydid" a separate cue to clap, which mimics the alternat-ing synchronicity of true katydids. The bat in the middle has to locate by sound that *original* katydid (again, "Katie"). He has a one in nine chance of doing so, and always fails. Then all nine kids are asked to clap in unison. The bat's chances of finding that original katydid among them are still one in nine. However, with the individual sounds of all of the clappers blended into one, the bat has more difficulty in homing in on any individual. In a real scenario, a possible advantage of truly synchronized calling could be the overwhelming of a bat's sonar.

While the bat's chances for a meal are increased in this scenario, the individual katydid's risk of being eaten is decreased—eightfold!

There is a bit of a risk that a large congregation of calling insects will attract more predators. This no doubt happens. However, the general predator-to-prey ratio in the animal kingdom usually allows for there to be enough of the latter to survive the appetites of the former.

Are these insects chorusing on purpose? Maybe; maybe not. A study by Michael Greenfield and Igor Roizen suggests that the chorusing of katydids can be a side result of male competition.[6] In experiments with Panamanian coneheads *(Neoconocephalus spiza),* they concluded that the insects, instead of intentionally joining one another in song, were instead attempting to jam each other's call. The females, who showed a marked preference for the lead singers in the group, caused the males to attempt to take that lead. The resulting synchronicity, the study suggests, did not evolve under selection pressure to generate a collective outcome, but instead evolved as the result of an individual's attempt to call attention to itself.

A similar study was undertaken with Indian bush crickets. With all the males attempting to be the lead-calling cricket, the male chirps were just milliseconds apart, which to our ears sounds as if they are blended as one.

One thing to note is that *Neoconocephalus spiza* of that area were found to have no "phonotactic" natural enemies, that is, no bats or wasps that target their calls. Here is a situation where the "safety in numbers" scenario appears moot and shows that there can be more than one trigger for a particular phenomenon.

Whether Ensifera are joined with others or going it alone, there can be no question that their song has a purpose—several, actually. It is part of their inheritance. They are born with their auditory signals encoded in their systems. A singing insect is compelled to sing. To silence it, you'd have to tie its wings behind its back. This becomes most evident to me on those sunny November late afternoons when I'm noticing how quiet things have become. The frosts have come and gone several times, a grim reaper harvesting the last of the year's insects. The chorus has been silenced. And then I hear a trill. It's a lone Carolina Ground Cricket *(Eunemobius carolinus)* calling, feebly, and stuttering, from beneath a leaf in the side yard. The song lacks the vitality of its summer brethren, but those worn wings still move at a blur. There are no females left to answer. It doesn't matter. It is trilling because it has to, and it is giving it everything it's got. It is the violinist playing as the *Titanic* is sinking. It's what they do.

How can that not stir a soul?

two

Why Listen?

Picture this: A comedian is on television. He stands on stage behind his microphone, plugging away at his repertoire. He tells another joke. No one laughs. The camera pans across the silent audience. Crickets are heard in the background.

The sound of crickets chirping has long been synonymous with bombed acts. It is meant to convey silence, the worse possible state for a performer attempting to elicit an enthusiastic response. I was thinking about the irony of this, that a sound is used to illustrate the *absence* of sound. Why not just have that actual silence? Does soundlessness seem less quiet to us than the call of crickets?

I think I know why this has become an entertainment industry practice. This device works because we *do* associate silence with chirping crickets. The song of the cricket is so ingrained in our consciousness, that once you erase the human din, you reveal that underlayment of sound.

The crickets most typically overdubbed for this effect are Snowy Tree Crickets *(Oecanthus fultoni)*. Sometimes they use field crickets. The reason, I am guessing, is that they are the culprits most noticeably (even though we don't always notice it) producing calls that dominate a quiet night. Add to that the fact that they maintain a steady rhythm of chirps, much like

a ticking clock marking the passage of time. The longer the
"clock" ticks away, leaving that performer squirming on the
stage, the more we can feel his or her discomfort, which is the
point of the scene.

I suppose it doesn't hurt that these crickets produce a call
pleasant to the ear, too. If you close your eyes and imagine a
warm summer evening, you need no other cue to hear them in
your mind's ear.

We all hear the crickets. We hear them every late summer
night. We hear them so much, that for long periods of time we
tune them out. Our brains are equipped with a self-regulating
aural-dampening mechanism to protect us from overstimu-
lation. If we couldn't turn off the noise, and even the most
pleasant sounds become noise when you've heard enough, we'd
go insane. Our hearing is not our only sense with an unconscious governor. Take our sense of sight. We see things we do

Snowy Tree Cricket *(Oecanthus fultoni).*

not realize we see. I am not just talking about our peripheral vision. Things right before our eyes go unnoticed. A perfect example of this occurred while I was walking along a wooded trail with my friend Bill. Out of the blue I said, "You know, I'd really like to see a Wood Turtle." Five seconds later, I nearly stepped on one right in the middle of the trail. Bill told me I wasted my wish on seeing a turtle, and next time I should conjure up a bag of diamonds. The turtle, however, was not the result of a fulfilled wish, as much as I'd like to believe that so. My theory is that I saw the turtle just prior to making that statement. While I hadn't *noticed* it, it was within my field of vision and, unbeknownst to me, the image of that wood turtle was planted in my brain. It's what made me wish to see one.

I missed seeing that turtle despite the fact I was looking for interesting things—like Wood Turtles—along the trail. This illustrates that no matter how hard you try, you can't see everything. Naturally, from person to person, there are varying degrees of acuity in observational skills. Bill is far better at noticing things than I am. In fact, it was he who pointed out the turtle, even though, technically (if my theory is correct), I saw it first. There are just some people who are more attuned to picking out aberrant details in the landscape. He told me, by the way, how he now curses me whenever he is walking across barren, gravelly areas. That is where you find tiger beetles. Tiger beetles are large-eyed, giant-mandibled frenetic hunters of other insects. They race after their prey like, well . . . tigers, chasing antelope. Prior to pointing them out to Bill, he was happy to take the opportunity to turn off those ever-chugging search engines when walking in these seemingly desolate, nonproductive areas. It does take a level of energy to maintain one's hypervigilance. Now, he says, "I can't even stop looking for stuff when I'm in the dirt!"

This phenomenon of turning off our awareness occurs with

all of our senses. Rest your hand on your leg, and your leg feels your hand there. The sense of touch in your hand feels your leg beneath. After a while, though, neither feels neither as your brain moves on to more interesting distractions. What does the room you're in smell like? Did you even notice before you read that sentence? Is there a particular taste in your mouth? If there wasn't before, you are probably noticing something now.

The point is, sources of stimuli within our reach do not cease to be because we do not notice them. They are there to be seen, heard, felt, smelled, and tasted when we are ready to do so. What we ultimately notice, however, is what we are interested in. I could walk side by side down a street with my wife, Betsy, and at the end of the road she would be able to tell me the styles and colors of all the houses we passed. She'd then further impress me by recalling the makes and models of the cars in the driveways. I would have to take her word for it. Unless I'm making a conscious effort to notice them, those things are virtually invisible to me. Even if I were trying to notice them, my mind would begin to wander. However, I could tell *her* about each of the birds I saw on the lawns and list the insects I heard calling. She would most likely have looked right through the birds, and the insect song would have just been part of the overall atmosphere.

It's a different story when it comes to *seeing* the insects making those songs I would have heard. They are either under or behind things, or so cryptically designed that noticing one is difficult even when they reside in the upper hierarchy of things you wish to see. With crickets and katydids, there's the added challenge of seeing creatures that are most active in the dark. We must therefore rely upon our ears. To truly experience what these insects can contribute to our aesthetic pleasure, we must listen.

The thing is, though, a majority of the time most of us have those aural dampeners I talked about engaged. Our autopilot selects which sounds to notice and which ones to ignore. While our ears are picking up the chorusing of the crickets and katydids outside of our windows, the brain is for the most part tuning them out. The sounds drift in and out of our consciousness, rising above, and then succumbing to, the ever-present, ignored, ambient noise.

Because of this, we miss out on a most extraordinary concert being performed from the trees, shrubs, and fields that surround us. These songs are not created for us to hear. They are not lighthearted ditties and jingles performed to delight a human audience. They are life-essential mechanisms employed by the Orthoptera in order to propagate their species. They are sounds to attract, and sounds to repel. So are the calls of Wood Thrushes and Spring Peepers, yet we take such delight in hearing them that we sometimes forget, in our anthropocentric world, that they have absolutely nothing to do with us.

That is a gift we humans have been given. We can take pleasure in hearing sounds. It doesn't always matter why that sound is being made, or who or what is making it. Sounds can enrich our lives.

I notice the sounds of crickets and katydids. I notice the sounds of singing birds and chorusing frogs. I hear caterpillar droppings bouncing off leaves in the summer, and the rustling in the understory along woodland edges. I notice these sounds more than most people do because I have consciously tuned in to them. I've been lending them my ear since I was a very young child. Unfortunately, having trained myself to notice sounds, that ability can be difficult to turn off. Being ever ready for a random sound to trigger my attention makes me more susceptible to hearing unpleasant sounds. My windows are open as I type this, and I hear an American Goldfinch, Great Crested Flycatcher, Scarlet Tanager, Wood Thrush, and Baltimore

Oriole outside my studio. I can't begin to tell you how much I enjoy those sounds! But I also hear the fan on my computer, and wish it were quieter. Jim, my rat terrier, is snoring on the couch across the room. Snoring goes right through me. A neighbor, about two houses away, is running some kind of loud piece of machinery. It's a dull drone, but I hear it, try as I may not to.

You have to take the good with the bad, though. I remember hearing a man with synesthesia being interviewed on the radio. People with this syndrome see sounds, hear or smell colors, and/or taste shapes. Somehow their brains cross the wires to connect these seemingly isolated senses. While this often provides the welcome bonus of added stimuli, it can also add unpleasant elements to what normally would not be perceived as such. When asked if he believed his syndrome was a gift or a curse, the man turned the question around on the interviewer, saying, "Think of how often you smell things unpleasant. Would you give up your sense of smell so you would never have to smell those things again?" The answer, of course, was no. That would be my answer, too. Hearing the occasional unpleasant sound will never trump the benefits in hearing the many more on the other end of the spectrum.

Incidentally, I became curious about how the natural sounds of night would be perceived by someone with synesthesia. An Internet search turned up a couple of examples. One person sees cricket chirps as the color red and frog croaks as blue. A child saw cricket chirps as a little *white* noise, while a woman saw them as a brown, flashing, balled-up fist. I tried to see if I could stimulate a synesthetic episode in myself. I played a recording of Spring Field Crickets, and closed my eyes, trying to think of nothing. The image of Spring Field Crickets, wings a blur, played in my mind. The color tan was evident, but it was provided by the dead leaves they were calling from. Technically, one sensory pathway merged with another, which is a quality of synesthesia. Sound merged with image, but I was trying for something less literal. A true synesthete would see the image as an overlay in their actual surroundings, as opposed to it being limited to what they knew of the source.

I should mention at this point that my hearing is only so-so. I've had it checked. Not only do I have some hearing loss in my left ear, but I also suffer from tinnitus, a ringing in the ears. What I hear is an ever-present high-pitched static, most noticeable when there are few other distracting sounds. Throughout the year I sleep with a fan running (the sound of which one synesthete sees as orange). The whirring creates a white noise that distracts my attention away from the whirring in my ears. Ringing in the ears is one noise I work hard to ignore, and for the most of the day I can. Of course, now that I'm writing about it . . .

So, how do you turn on, or up, the sounds of the crickets and katydids? How do you allow them within that upper hierarchy of what you choose to notice? How do you tune in to the singers of that giant "cricket radio" outdoors? The answer is something I have been exploring for a number of years, and it began with learning birdcalls.

As a birder, I have enjoyed the songs of birds as much as I've enjoyed the experience of seeing them through a pair of binoculars. I began this pastime with an interest in finding new species, which meant laying my eyes upon them. It soon became evident, however, that I could find and identify new species by hearing their song. Each bird species has a distinctive call. If you learn the song, you know who's doing the singing. It's called "birding by ear."

Sometime around the mid 1980s I purchased the *Birding by Ear* cassettes, narrated by naturalist Dick Walton. I listened to them when I worked, or would pop them in the cassette player to bone up on my calls while driving. I'll never forget the excitement of identifying for the first time, based solely upon hearing its call, a bird I'd never seen before. I was out in my yard and heard from the bushes, "Git-ti-beeeeer-chk— git-ti-beeeeer-chk . . ." I remembered that call on the tape, particularly because of the "handle" it was given. A handle is a phrase that helps connect a call with words. It's also known as a mnemonic. The handle for "Git-ti-beeeeer-chk" was "Get the beer check." I said to myself, "That sounds like a White-eyed Vireo!" I ran in to the house, grabbed my binoculars and then located the bird. It *was* a White-eyed Vireo. I watched as it sat atop an autumn olive, repeating its call over and over. The image of the bird merged with the sound it made. Although I had watched birds call before, this was the first time the sound a bird made took precedence over the enjoyment of taking it in visually. It was the sound that led me to it.

From then on, I made a point to linger on the sight of a bird until it called, or to try to absorb the image of a bird I heard singing. I wanted to watch that bill go up and down while the call escaped its pharynx. I found this to be the best way to learn birding by ear. Within my brain, the sound became associated with the image in that quasi-synesthetic way.

I soon discovered there are other sounds out there to be heard. My true awakening to those made by insects came about after reading Vincent G. Dethier's *Crickets and Katydids, Concerts and Solos*. Prior to reading this classic book on Dethier's quest to learn more about the calls of Orthoptera, my ability to separate one insect call from another was limited. I knew the basics: Spring and Fall Field Crickets, Snowy Tree Crickets, Common True Katydids, and some of the grasshoppers and meadow katydids. I had learned them by identifying them by sight, and then hearing them call. But Dethier provided a way to approach these bugs from a different direction—sound. In addition to the engaging prose describing his hunt for these creatures, he included at the back of the book a phonetic key to the calls of thirty-six species. Soon I was out in the field, book in hand, looking to match the sounds I heard with the descriptions in his book. My first victory was the ubiquitous Fork-tailed Bush Katydid *(Scudderia furcata)*. There can be no doubt that I have heard them call all my life. This is evidenced by the great numbers of them I've since discovered in places I frequent. Their calls, although ignored, undoubtedly entered my ears. Ears are always open, regardless of whether or not their intake is being consciously processed. It wasn't until I was out listening for those katydid calls on a muggy August night that I actually *noticed* them. They were all along the edges of my yard! What I heard were short, random, soft "tsips," coming from the lower vegetation. Dethier's book describes the call as "a soft lisping zeep or zeep-zeep-zeep at irregular intervals." He wrote that they would be found in "bushes and tall grasses bordering woods." The next step for me was to find one, which wasn't that difficult. I traced the call to a little green "leaf" perched atop a blueberry bush. It didn't seem overly phased

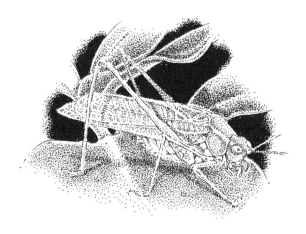

Fork-tailed Bush Katydid *(Scudderia furcata).*

by my flashlight, and soon resumed its calling. The wings vibrated in the blink of an eye—"tsip . . . tsip . . . tsip . . ."—in about fifteen-second intervals. In the background its brethren (competition, actually) returned the aural volley. This was the first time I watched the actual sound emanate from a katydid's wings. The connection was made. I was hooked. Now when they call, I cannot *not* hear them. I cannot *not* see them, either. A picture comes to mind of the diminutive leaf mimic making that sound from the dark edges of the woods. It is a gift for two of my senses; the sense that allows me to enjoy the aesthetics of a thing that is pleasing to the mind's eye, and that which creates a gentle sound to tickle my eardrums.

Occasionally, while doing a lecture or leading a trip, I'm asked the question, "What good are they?" This would refer to the subject, be it a moth, frog, katydid, or, very rarely, bird. For some reason the public finds it easier to embrace birds. The question is usually not meant to be derogatory about that subject. I believe the person asking is searching in earnest to find some value in it. People tend to need to do that. To truly appreciate something, they need to perceive a quality in it that is beneficial to humans. I often stumble at this point, because I feel the need to search for an answer that will satisfy the asker. Rarely is that answer one that would answer that question had *I* raised it. For example, having written a book on the topic of moths, I am often asked about their value to us. My go-to answer is always: "Moths are valuable because they pollinate a large percentage of our fauna."

Sure, that's good and all—we certainly need our plants pollinated. However, I really don't care so much about that. I'm just glad I can tell people that fact so they will stop squashing moths on sight.

I once served on a local Inland Wetland and Watercourses Commission, which is charged with protecting vernal pools, a habitat Spotted Salamanders need to reproduce. The worth of the salamanders was occasionally put to the question. The answer had to place a value on these amphibians, creatures some people couldn't, to quote one commissioner, "give a tinker's cuss about." So, when asked of their value, I'd answer, "Spotted Salamanders are valuable because their larvae feed on the larvae of disease-spreading mosquitoes." But here's my own reason for wanting to protect them: There is a certain primitive nature to them. I see it in their form and in those cold, black eyes. I see these creatures and sense the long journey of the species from the primordial soup of the Carboniferous period to the dark, cold, rainy nights of their mass migrations. They drag themselves across the land, but rediscover their grace in the watery habitat from which they sprung. How can I rest easy in destroying a population that has traveled so far, when there are alternatives that only cost money?

Try using that argument to convince a developer to spend tens of thousands of dollars to circumvent the road around a vernal pool buffer area. Or to convince that "non-tinker's-cuss-giving" guy to vote in favor of rerouting that road. Raise the question of public health, and you have a stronger argument.

Here is a question I was asked recently while doing a radio interview on the topic of night-singing insects: "What *else* do the crickets and katydids tell us?" The interviewer had asked this after I had described how you could gauge the temperature by counting the chirps of a Snowy Tree Cricket *(Oecanthus fultoni)*. I stumbled on this question, too. It's good this was being prerecorded, because I was breaking the cardinal rule of radio in filling too much time with dead air. Fortunately, fellow amateur orthopterist Michael DiGiorgio was with me and came to the rescue. He said that their presence could show the health of a habitat. An area rich in singing insects suggests a healthy ecosystem, and, he added, the calls of individual species could be a clue to the type of habitat.

Relieved he filled in that blank, I gratefully moved on to another topic. I thought about this again when I got home, though. Actually, the question was dogging me. Why couldn't I answer it? At first I thought it was kind of a silly question, asked in the vein of "What good are they?" Mike's answer was perfect in that regard, even though I know he too places a value on these insects that transcend their role as indicators of environmental health. The more I thought about her query, the more I appreciated its depth. It was actually a very thoughtful question. What *do* the crickets and katydids tell us? What do they tell *me*?

As it turns out, they tell me very much, and I've discovered many reasons for listening:

- The chirping of Spring Field Crickets *(Gryllus veletis)* tells me that the cold of winter is behind us. There are many false starts to spring in New England. Flowers bloom, peepers peep, stoneflies hover over the streams and ponds. Then, just as you're starting to put the flannel shirts away, you get hit with a foot and a half of snow. It's just like in those horror movies, where the monster is vanquished and the soft music starts playing. Then "ROAR!" the monster jumps out at you! Once the Spring Field Crickets begin to call, though, there's no backpedaling to winter. Winter has been vanquished, for real this time. It's onward to summer.

- The steady shushing of Eastern Swordbearer *(Neoconocephalus ensiger)* coneheads tell me we are in the thick of summer. While there are a number of other singers out there signaling the same, these coneheads seem best able to flick on that switch of recognition. I suppose part of the reason for this is that when I hear them, it is always unexpected. Because they do not call in my yard, or anywhere in my immediate vicinity, I first hear them while driving with the windows down, my mind preoccupied with any of an infinite number of things besides coneheads. I hear them and it's always, "Wow, they're calling already? Wasn't it just spring?"

• Conveying the passage of time seems to be a particular talent of the crickets and katydids. When I hear the late daytime calls of Common True Katydids *(Pterophylla camellifolia),* I know that in a few weeks I will look out in the morning at a silvery, frost-covered lawn. As the light cycles shorten and their time on this earth nears the end, the insects push the envelope to find mates and lay eggs. For the true katydids, this means opening the shop earlier.

A different insect, the Striped Ground Cricket *(Allonemobius fasciatus),* presaged winter for the late entomologist Harry Piers:

> The sound has a peculiar silvery timbre, and when myriads are shrilling all over the fields at night, or on fine days in late autumn, when other sounds are hushed and the air filled with the mystic charm of the hour or season, it produces a peculiarly drowsy, cease-less tremor, pulsation or 'shimmer' of sound which is very familiar and loved by all dwellers in the country. It is, however, ineffably associated with a sad feeling that summer is on the wane or past, for it is our most characteristic autumnal sound in grassy places, and is linked with the sight of golden-rods and purple asters and the odour of falling leaves.[1]

They don't do that for me. They tell me something else, though. As Mike said, you can tell something about a habitat

Eastern Swordbearer *(Neoconocephalus ensiger).*

by what you hear calling. For years I had associated Striped Ground Crickets with wetlands. Their "chit-chit-chit" call is ever-present in marshes and fens and along the edges of rivers and lakes. I then came across a population in a friend's yard. They were the dominant species, and numerous. The area was quite dry, and was used as a riding ring for horses. When I looked more closely at the ground, I noticed it was sand, which was trucked in for the ring. It occurred to me, then, that they prefer the well-drained sand substrate, not so much the proximity to wetlands. Now when I hear Striped Ground Crickets calling, they are telling me about the ground upon which I walk. It is likely to be sandy.

- The call of the Northern Mole Cricket *(Neocurtilla hexadactyla)* tells me there is water nearby. They build their network of burrows in the damp mud and sand along the margins of streams and ponds.
- One summer afternoon, I was hiking along a trail. Dusk was approaching, and at the very edge of my hearing I heard a faint trilling. It was the combined call of many insects far in the distance. The trilling grew louder as I drew nearer. I recognized the calls as those of Pine Tree Crickets *(Oecanthus pini)*. While I could not see any conifers where I was, I knew I'd be approaching them. The trail opened up, and there, along the edge, was a great stand of eastern white pines. The calling Pine Tree Crickets told me those trees would be there, well before I could have seen them for myself.
- The whistling chirps of Jumping Bush Crickets *(Orocharis saltator)* on a bustling city street in Manhattan reminds me of the resiliency of nature. If a tree is going to go through the trouble of growing in a noisy, crowded, urban setting, there are likely to be birds and insects to

complement it. That a singing bush cricket will declare this tree its own and call out to invite females to join him shows that single-minded tenacity in range expansion. It is what allows all manner of biota to extend to their limits. I have to assume that those metropolitan males are successful in luring mates, and that new generations of Jumping Tree Crickets will call those trees home.

- Hearing House Crickets *(Acheta domesticus)* sing in a pet shop reminds me of how easily we humans choose sides. We both collectively and as individuals, and seemingly at random, assign levels of value to different forms of life. Given the sound of those cricket chirps, we might be forgiven for assigning it a "cheery" quality. And yet it is the lot of those singing males to be sucked into the gaping maws of frogs, lizards, and snakes. That was the fate of their parents, and will be the fate of their offspring. House crickets, wild Gryllidae in their native Asian and European lands, have long been relegated to pet food in the United States. We've chosen to favor the herpetofauna over the Orthoptera. That's a tough break for the American House Crickets.

In China, however, it is the other way around. Crickets can be infinitely more valuable than a pet reptile or amphibian. Not only are they cherished for their song, but a good fighter can bring in thousands of dollars.

- Silenced song tells me something, too. I was listening to two Common True Katydids calling up in the oaks along a New Jersey trail. They suddenly stopped in midcall. I was already shining my flashlight into the upper canopy, searching for them. At first I thought my light had scared them, but that had never happened before. Finding a true katydid from the ground is a tall order—not that I'll ever stop trying. Then a Large Brown Bat flew into the beam. While it did not appear to have caught one of the katydids, I believe its presence forced their silence.

There are times when I hear katydid calls come to a sudden stop. Sometimes, for example, the wind forces them to concentrate on hanging on to their leaf. Now when they stop, and there is no wind to blame, I add bats to the list of possibilities.

In mid July there is a cricket silence that lasts for a week or two in my own backyard. This tells me the adult Spring Field Crickets are dead. It also tells me that the exoskeletons encasing the Fall Field Crickets are getting a little snug. Within a few days they will molt for the last time and pick up the song where the Spring Field Crickets left off. That silence forms a symbolic bridge from spring to fall.

- It should come as no surprise that the Ensifera are giving us audible clues to their whereabouts. What I am continually reminded of, however, is how much I should expect the unexpected. In 2007 I was recording insect songs on Connecticut Water Company property in Killingworth, Connecticut. Davis's Tree Crickets *(Oecanthus exclamationis)* and Carolina Ground Crickets were trilling away, and I managed to get some good recordings. I finished at around 4 p.m. and went home to load the calls into my computer. I heard something in the background of the recorded calls that I had not noticed while afield. It was a series of short, gentle buzzes. I listened to a few CDs of calls I had, and then went onto the Internet, where orthopterist Thomas Walker has recordings of most of the North American species. It sounded exactly like a Woodland Meadow Katydid *(Conocephalus nemoralis)*. The nearest record of this species was in southern New York, a few hundred miles southwest of Connecticut. I immediately went back to confirm what I thought I had. Not only did I relocate what were indeed, Woodland Meadow Katydids, but discovered I had driven past another colony of them on the way. Even though I neglected to hear them in the field, they were demanding to be noticed, whispering "Psst, we're here" in my microphone.

The call of another species not known to be in Connecticut led to its discovery. The nearest known populations of Black-legged Meadow Katydids *(Orchelimum nigripes)* were in Maryland and Pennsylvania, and yet there they were, ticking away in Hurd Park, East Hampton. Two years later I heard them in another part of the state, buzzing in an oak tree. These are insects that inhabit grasses and sedges, not deciduous trees, or so I thought. What was this telling me? It was dark out, but when I shone my light in the adjacent field, I got my answer. The meadow had been mowed earlier that afternoon. The katydids were driven out of the grass and into the trees. This was interesting, since Orchelimum generally escape danger by staying put. Certain individuals in this small population had a better idea. I would imagine they fared better than those that went down with the ship.

- Sometimes the singing insects tell me nothing I don't already know. Instead they reinforce, and round out, an existing perception. I hear their songs and am glad that the woods and fields are healthy enough to host the singers. To me, bug-eaten leaves are a welcome sight. To hear a mixed variety of species calling in one area is a confirmation of natural diversity. There are plants to feed and hide them. They, in turn, are there as food for other predators.

Black-legged Meadow Katydid (*Orchelimum nigripes*).

As I alluded to before, their expected songs can also serve as seasonal fence posts. As the seasons progress, I anticipate certain sounds to be in certain places. There is one place I go where, year after year, I can hear Curve-tailed Bush Katydids *(Scudderia curvicauda)* in the highbush blueberry shrubs. There is bog I can count on to hear Sphagnum Ground Crickets *(Neonemobius palustris)*. I have a handful of favorite spots for Northern Mole Crickets and Sand Field Crickets *(Gryllus firmus)* later in the season. Hearing them where they belong, when they belong there, is an audible affirmation of that time and place. I don't need them to tell me this, but they round out the experience.

Standing outdoors on a chilly November night, I am fully aware that the temperature is nearing the mid to low forties. I feel it on my skin. I see it in the condensation of my breath. But I can hear it, too. The Jumping Bush Crickets, Snowy Tree Crickets, and Common True Katydids up in the trees are calling so slowly, their songs are now more pause than sound. Their pitch has dropped, like the hoarse voice of a person struggling to speak through a sore throat. In addition to feeling (and seeing) the cold, I am hearing it through its effect on the stridulatory mechanisms on the insect singers. I welcome the autumn, despite the fact that the forest and field will soon be silent. It is my favorite season. I like the way it smells. I like the feel of the crisp, clean air. I know it's somewhat macabre to look forward to a season ushered in by those creaking, soon-to-be-dead harbingers, but I don't see it that way. To me, it's nature clearing the slate for another go at it. The work of those insects is nearly over. They did what they had to do.

I wish I could have done that radio interview over again. I'd still be talking, and the host would be sorry she ever asked the question.

three

The Straight-winged Bearers of Swords

Few people have heard the word *Ensifera*. The label describes a suborder of insect songsters that add a tuneful ambiance to a warm, summer evening. Even the sound of the word *Ensifera* is pleasant to the ears. It rolls off the tongue and is less of a mouthful than "katydids and crickets" or "night-singing insects."

The Ensifera, which are the focus of this book, are part of the order Orthoptera. We've heard the first part of that word before. When you need your teeth straightened, you go to an *ortho*dontist. To get your vision straightened, you see an *ortho*ptist. As you may have gleaned, the Greek prefix *ortho* means "straight." The *ptera* in "Orthoptera" means "wings." The "straight wings" referenced in this name calls attention to the somewhat parallel edges of the wings of the insects in this group. For the layperson, this may not seem the most obvious attribute, just as calling butterflies and moths "scaly wings" (the translation of their order Lepidoptera) may not point to what we first notice in those insects. However, if you observed a hundred butterflies and moths and were boggled by the many different sizes, shapes, colors, habits, and forms, you would notice that upon handling *each* of them, there'd be dusty little scales left on your fingertips. Those scales unite that diverse group. With the crickets and katydids, in which there is also a wide range of differences between the groups, you'd see most have in common those parallel wings. I say *most,* because there are some variations on the theme.

When one delves into the origins of names, there are many "Aha!" moments. As a person whose job it is to string words together, I may be a bit partial to the enjoyment of discovering where those words came from, so I hope you'll bear with me as we explore a bit of etymology.

The Orthoptera encompass two suborders, the Caelifera (grasshoppers) and aforementioned Ensifera (katydids and crickets). *Caelifera* is Latin for "chisel-bearing," and references the female's stout ovipositor (egg-laying apparatus). As with the parallel wings, it's not what jumps out at you when you look at a grasshopper.

Ensifera translates as "sword-bearing," again pertaining to the female ovipositor, which can be long, pointed, and laterally flattened. Historically, roaches and walkingsticks were included in this group. Some scientists place them, along with termites, mantids, and earwigs, in an even larger "Orthopteroid" category. Orthopteroid can be thought of as Orthopter*ish,* meaning they share many of the Orthoptera features. These traits include:

- Wings that can be folded over the top of the body
- Growth through incomplete metamorphosis
- Chewing mouthparts
- The presence of a pair of appendages (cerci) on the tip of the abdomen.

Today most entomologists consider the Caelifera and Ensifera to be the sole members of the Orthoptera order. Fossilized records of these insects date back to the late Permian period over 250 million years ago. By the time modern humans came into the picture, the Ensifera were long on hand to regale us with their song, and the Caelifera (some of them, at least) were itching to get at our crops. They are found all over the world, wherever there are terrestrial plants, the exception being areas that remain frozen year round. Over the millennia they've hopped this earth, they have evolved and split into over 24,000 species.

Wings Make the Orthoptera

I think it's safe to say that the outer wings of these insects make them what they are, beyond the fact that their straight edges lump them together within their order. The Caelifera and Ensifera have two pairs of wings. That outer pair is called the *tegmina* (singular is *tegmen*). They contain the instruments with which the insects call, and they help them blend in with their habitat. These structures are thickened and hardened, sometimes leathery. They also protect the more delicate pair of hind wings beneath, much like a beetle's curved outer wings (elytra) protect *its* hind wings. In species that fly or glide (not all of them do), it is the pair of hind wings that carry them. Some of the Orthoptera have greatly reduced tegmina and hind wings, or none at all. While this may ruin that whole "straight wing" theme, we must remember that there isn't a generalization out there that's free from exceptions (probably including the one I just made).

The Head

The head, as heads do, contains the mouthparts and most of the sensory structures. The antennae emanate from the top and are used as tactical receptors to collect information on the surrounding environment, and other Orthoptera. I remember calling them "feelers" when I was a child. I'm not sure where I'd picked that up, but it turned out to be a good description of their function. They play a similar role to whiskers on a cat's face, or our fingers, which when swept over an object can feed information to our brain. I suppose the term *feelers* was a comfortable one, since it's so obvious that's how they were being used.

In the tree crickets, patterns on the base of the antennae can be used to identify species. Nearly all of the North American species have a unique pairing of spots.

Paired appendages, called palpi or palps, are located along the sides of the mandibles. They are used for "smelling," tasting, feeling, and manipulating food.

Close-up of Narrow-winged Tree Cricket
(*Oecanthus niveus*) antennae base markings.

The mouthparts of the insects in this order are designed for chewing and shearing. Most Orthoptera chew plants, but many of the Ensifera supplement their diet with a meal of other arthropods, detritus, and fungi.

The Power Station

The power station of an insect is found in the midsection, or *thorax*. It houses the flight muscles and muscles to work the legs. Those legs are composed of three main sections: The *femur* (pl. *femora*), sometimes referred to as the "drumstick," is the heaviest segment, and is attached to the lower area of the thorax. The *tibia* (pl. *tibiae*) fills the middle section, and could be compared with our own forearms and shins/calves. The hind tibiae often have spikes, which can be used by the grasshoppers as a method of defense. Ground crickets have movable spurs on their hind tibiae, which are chewed on by females during mating. The spurs secrete a fluid that is lapped up by the female. The *tarsus* (pl. *tarsi*) is composed of three to four sections, forming the "foot."

In the Orthoptera, the upper (dorsal) surface of the thorax is fused to form a kind of shield that wraps around the sides of the thorax. This is the *pronotum,* and the shape can sometimes be used to tell one species from another. In some species, the pronotum helps resonate the call.

The Bulky Nethers

The posterior segment, or *abdomen,* is the bulkiest part in most insects. It contains the reproductive, respiratory, and digestive organs. At the very tip of many orthopteran abdomens, are paired appendages known as *cerci* (sing. *cercus*). The varying shapes and lengths of these apparatus can often be used to sort out species. The earwig is an example of a familiar insect with prominent cerci. You can't hear "earwig" without thinking of two things: One, *pincers,* which are the cerci of this Orthopteroid insect. And, two, well, creatures that crawl into a sleeping person's ear to lay eggs (which then hatch and drive the victim insane as they eat away at the brain). Granted, with regard to the latter, because of their preference of dark, moist places, an earwig *could* conceivably crawl into someone's ear. But they don't, and to quote an unlikely name (Martha Stewart) in a book about insects, "And that's a good thing."

In the case of the cerci for these insects, the pincers serve the function of an extra pair of fingers. They are used for defense, grooming, courtship, and realigning the wings. However, in the Ensifera their usage is less dynamic. They are sensory organs, and in some species they may be used to align the male in a better position for mating.

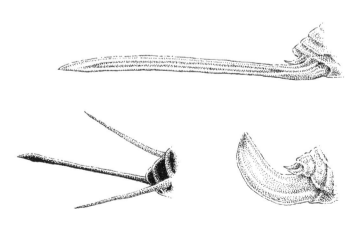

Conehead, cricket, and false katydid ovipositors.

The male sexual organ, *aedeagus,* is internal and cannot be viewed with the naked eye. The female egg-laying apparatus, called the *ovipositor,* is external and can often be used as an identification aid. Depending on the family, the ovipositor deposits the eggs on or in plants, rotting wood, or soil.

Robots versus Leaves

Although this book focuses on the Ensifera, it would be helpful to know what separates them from the Caelifera. Grasshoppers, as they are commonly known, are one of the first insects a child can pin a name on. The more familiar representatives are usually species found within the Acrididae family. These look like what most people think of when they hear the word *grasshopper.* I've always thought that grasshoppers resembled tiny robots. There's a somewhat mechanical look to their design. Katydids, which are morphologically similar, have more of a leafy, organic look. A twenty-foot grasshopper would most likely be the subject of a science fiction movie. A similarly sized katydid would fit more in the fantasy genre. The katydid, though, is the one you'd want to run away from, because one that size would be more likely to eat you. The giant grasshoppers would be less interested in you than in the trees you were hiding behind as it approached.

I'll admit stretching the comparison a bit, especially because most of the Caelifera have evolved into shapes to blend in with grasses and stems and bear patterns that add to that cryptic effect.

Despite this, we do notice grasshoppers. Any time you walk through a meadow on a summer afternoon, you are bound to see them as they hop or fly out of your way. It is that action that grabs our attention. Most in this group are found on the low-lying plants upon which they feed, and typically they *walk,* not hop, from one place to another.

It should be noted that the word *locust* is usually reserved for migratory grasshoppers. These are in the Acrididae family and are known for their swarming behavior.

Naturally, the first feature that stands out in a grasshopper's appearance is those long hind legs. Most of the Orthoptera share this feature. The legs are used to propel the insect from danger in a quick burst of speed, powered by those ample femora. Jumping can also provide a shortcut from one plant to another. The act of jumping expends extra energy, and to escape a perceived threat, a fleeing grasshopper will typically hop as few times as it can get away with. Some will fly out of harm's reach, but again only the minimal distance needed to bring it to safety. In addition to the energy burned in the act of hopping or flying, there is the added danger of exposure to predators. It is easy to imagine this risk if you think about your own discoveries of these insects in the field. Walking through a meadow, you tend to notice the grasshoppers (and smaller meadow katydids) jumping and flying upon your approach, rather than those sitting tight, hugging the stems of the surrounding plants. If you were a hungry bird, the jumpers would more likely end up as your meal. In fact some birds, such as American Kestrels, American Coot, and turkeys have been observed flushing grasshoppers, forcing them to reveal themselves by jumping.

Some grasshoppers do hop more than others. For example, there is a winsome group of little toadlike Caelifera in the family Tetrigidae. They are known as pygmy grasshoppers, groundhoppers, or grouse locusts. The late entomologist Harry Allard wrote that they "leap almost as actively as fleas." Their tiny size and mottled, dead-leaf look probably buy them a little more safety than the larger species enjoy.

I wrote about the hopping of frogs in my book *Discovering Amphibians: Frogs and Salamanders of the Northeast.*[1] Frogs, like grasshoppers, are thought of as *hopping* creatures, and in this case that perception would be more accurate. I posited that for a frog, hopping is most likely safer than crawling. A steadily crawling frog would be giving a predator more time to locate it. When a frog jumps, then stops, jumps, then stops, there are longer moments when it is motionless. This could increase the chances that a predator would give up looking for what *was* moving and turn its attention to something that *is* moving.

However, to find food, frogs need to move around considerably more than a grasshopper or katydid. The Orthoptera spend most of their time *on* their food and can afford to creep around on their plants. Most are not designed for long series of sustained hops. They don't need to be.

While katydids and crickets share those long hind legs with the grasshoppers, they posses a feature that readily separates them from their cousins: long antennae. The antennae of the Ensifera are generally longer than their body. The antennae of the Caelifera are considerably shorter than the body. This led to their nicknames, *long-horned* and *short-horned grasshoppers.* It is likely these differences reflect their preferred hours of activity. Grasshoppers are sun lovers. They have well-developed eyes, suited to finding their away around in bright, open meadows. A katydid or cricket tends to be more of a night owl. It is true that many of them are active in the day, but even then, they are more creatures of the shadows and leafy understories. Most of those active in the day will continue to be active once the sun goes down. Having enhanced sensory apparatus independent of available light levels can help them feel their way around, and find one another. This adaptation can be seen in insects dwelling in dark caves. They often have abnormally long antennae when compared with their "outdoor" counterparts.

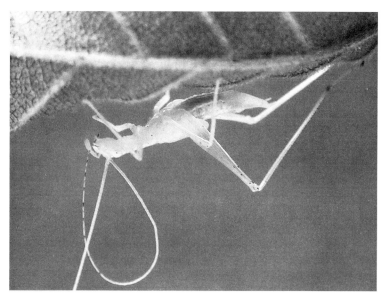

Preening Narrow-winged Tree Cricket *(Oecanthus niveus)* nymph.

An endearing trait of the Ensifera is their fastidiousness. They spend a great deal of time grooming. Katydids and crickets can be seen drawing the entire length of an antenna through their mouthparts. This removes debris that accumulates in the course of their wandering, and is usually done after a meal. They also use those mouthparts to wash their "feet," or tarsi. This act has a very *felinesque* look to it.

There are a number of katydids similar in shape to grasshoppers, most notably the meadow katydids, coneheads, and shield-bearers. The average person, not being aware of the differences in antennae length between the two suborders, would likely call a meadow katydid or conehead a grasshopper. Females, however, have another obvious feature that separates the two groups. The ovipositors of grasshoppers are short. They are used to insert foamy masses of eggs into the soil. Some species will deposit them in plants or rotting branches, but this is not typical of the group as a whole. Ensifera ovipositors tend to be quite a bit longer in comparison. This feature makes it very simple to separate those katydids from a grasshopper. They have long, laterally flattened ovipositors. In some species, they are longer than the body!

Stridulatin', Crepitatin'—Makin' Noise to Call a Mate'n

Two more differences between the two suborders involve their methods of producing and hearing sound. While some grasshoppers, like the Tetrigidae I mentioned earlier, do not produce sound, most of the others do. One type of call, known as *stridulation,* is created by rubbing the inner surface of the femur against the lower, outer edge of the tegmen. This produces a scraping or lisping call. It is a dry sound, similar to that of two sheets of sandpaper being rubbed together.

The other kind of call, employed by the band-winged grasshoppers, is known as *crepitation.* This is produced in flight and is the result of the rapid opening of the hind wings. The effect is like the quick snapping open of a Chinese fan. In the grasshoppers, this creates a crackling sound, which is often accompanied by a hovering, or zigzagging, flight, and a display of boldly patterned wings.

The Ensifera also create percussive sounds, but to do so they utilize different parts of their anatomy. Located at the base of the tegmina is the *stridulary organ,* also known as the stridulary area. This is composed of the *file* and the *scraper.* The file is located on the underside of the upper tegmen. While the teeth of the file cannot be seen from the top of the wing, the stridulary vein, which houses the teeth below, is often apparent as a raised, oblong area. When the wings are closed from the open position, the teeth of the file rub against the upward-pointing scraper on the lower tegmen. There is an area next to that scraper called the *mirror*. It resembles the skin of a drum stretched tight around the hoop, and serves the same purpose— resonation. The sound produced is modified by the speed at which the wings are opened and closed, the length of that wing stroke, and the shape, size, and venation of the tegmina. The configuration of the adjacent pronotum can also play a role in shaping the sound. Because the wings do not produce sound when opened, the insects cannot produce a continuous call. What we hear are combinations, or *trains,* of pulses. A pulse is the sound made when the tegmina close one time. A trill, which to us may sound continuous, is created by the very rapid repetition of opening and closing the wings. A sonogram of a slowed-down call will show each individual pulse that was linked to create that trill.

I have a small collection of pinned specimens of katydids and crickets. I've never been able to bring myself to send them to the great beyond, but some have died in captivity for a variety of reasons. Curious to see if a "dismantled" katydid corpse could call from the grave, I removed the wings of some of them. By rubbing them together I experienced moderate success. I couldn't come close to mimicking the more complex calls, but I could do a fairly decent impression of a Fork-tailed Bush Katydid, which gives a shorter "tsip" call.

The calls of the Caelifera take place during the day. There are those among the Ensifera that call during the day, but many of those continue into the night, and most are either primarily or solely nocturnal singers.

The last noticeable difference between the two orders is where their ears are located. The hearing organs, or *tympana,* of grasshoppers are found on the sides of the first segment of the abdomen. In katydids and crickets, they are located on the forelegs at the base of the tibiae. They can appear as slightly

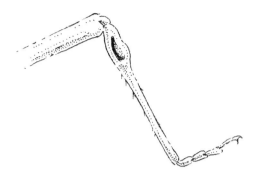

Tympanum on front tibia.

pinched areas with a little indentation in the center. Some of
the Ensifera, like the camel crickets, lack them entirely, just as
some lack the organs to produce sound. The ability to produce
sound, however, cannot always be gauged by what the human
ear can pick up. As I've mentioned, some Orthoptera create
ultrasonic calls that fall beyond our range of hearing.

Bigheaded Babies

The Ensifera begin life as an egg. Or does life begin as the adult
insect? Regardless, the eggs are laid collectively in the ground,
stems of plants, or rotting logs, or are attached to leaves. Most
eggs hatch in the spring or early summer, having overwintered
in this stage. Some can take a couple of years to hatch. The
season of hatching and the number of generations per year can
depend upon what part of the world they are in. In areas where
the climate stays warmer for longer periods of time, there are
species known as *bivoltine,* meaning there are two generations
in a single year. Warmer winters also allow a number of species
to pass the season as a nymph or adult. However, the more typi-

cal scenario in North America is that of the *univoltine* insect, where there is just a single generation in one year; the adults of most species die by winter.

From that egg emerges the first instar nymph. The Orthoptera progress to the adult stage through simple, or incomplete, metamorphosis. There is no larval or pupal stage. The nymphs are instead, miniature, flightless, silent, and sterile forms of the adult. As they progress through the typical five or six molts (some more) it takes to reach adulthood, they grow larger and the wing buds slowly take on the shape of the adult wings. As with the larvae of other insects, it is the job of the nymph to take in nourishment to bring them to the next stage. They are also taking on the task of dispersal, although to a lesser degree than the adults.

What always stand out to me on the katydid nymphs are their disproportionately long hind legs. They bring to mind the big ungainly paws of puppies on their way to becoming big dogs. As for the cricket nymphs, their heads look way too big for their bodies. Stephen J. Gould wrote about neoteny, which is the retention of juvenile traits in adults.[2] He observed that human babies have large heads and eyes in proportion to their bodies, and human adults are wired to find that "cute." Disney capitalized on this aspect of human nature by giving large heads and eyes to his cartoon characters. This is not lost on me when I come upon an early instar cricket.

It can be difficult to determine the species of early instars, but at this stage they are fairly easy to rear out. As they eat, they grow, eventually becoming too large to fit within their rather inflexible skin. It is at this point they molt. The skin splits, the soft insect squeezes out, and, as many do, may proceed to make a meal of its former attire. The newly emerged insect needs to lay low while its exoskeleton hardens. The meal offered by its shed skin offers a few advantages. For one, the insect does not

Round-headed Katydid (Amblycorypha) nymph.

have to travel far for its first meal. The skin also provides some instant nutritional value, and eating the skin removes evidence of the insect's presence from that immediate area. The last thing a young katydid instar needs is to offer to potential predators a clue to its whereabouts.

Once the insect reaches its last molt, it is just a few days away from being a fully functioning singing, flying, reproducing adult.

The preceding life history of a typical Ensifera glosses over the many traits and activities exhibited by the different families of katydids and crickets. Because I will be talking more about specific families and species, in the next two chapters I will break them down further into the three families of singing insects with which we most frequently cross paths: Tettigoniidae (Katydids), Gryllotalpidae (Mole Crickets), Gryllidae (True Crickets).

four

The Katydids

Family Tettigoniidae (Katydids)

With seven native subfamilies (plus one introduced), and over 275 species, the katydids are well represented in North America. They tend to be the colors and shapes of deciduous leaves and blades of grasses and sedges. Katydids are best told from crickets by a number of physical traits:

Katydids	Crickets
Four-segmented tarsi	Three-segmented tarsi
Left tegmen overlaps right tegmen	Right tegmen overlaps left tegmen
Wings held tentlike against body	Wings rest on the dorsal surface of body
Laterally flattened ovipositor	Needle-like ovipositor

The males generally call from perches, either as individuals or joined in chorus with other males. The females are drawn to the call of the male; many answer them back, and will determine, using some kind of katydid judgment system, whether or not a particular male measures up to their standards. As the female approaches, the male stops singing and may vibrate his body. This vibration is another method of communication, used to help the female find him. It's hard not to anthropomorphize this action as excitement at getting noticed by a potential mate.

The mating ritual begins with antennae contact. Should the male pass the test, the female goes behind him and palpates his abdomen. She then mounts him (in grasshoppers, this position is reversed). The male katydid has more than a song to offer his mate. First he attaches a sperm packet, known as a *spermatophore,* from the tip of his abdomen to the tip of hers. He then secretes from his abdomen a frothy packet known as the *spermatophylax,* which is then placed over that spermatophore. They separate, he bids her a sweet adieu, and she proceeds to devour the spermatophylax. This treat serves to buy some time for the spermatophore's sperm to enter her *spermatheca,* the chamber that stores the sperm for fertilization. There is also the possibility that without this nuptial treat keeping her preoccupied, she might remove the spermatophore herself before it had a chance to inseminate her. Finally, a female katydid likely receives some nutritional value in the offering.

There are numerous rewards to seeking out the seven subfamilies of katydids. They cover many different habitats, giving the curious naturalist even more excuses to get out among them. I often view the objects of my hunts, whether they are birds, bugs, or plants, as portals to the outdoors. They are what get me out of the house.

Bump-headed Meadow Katydids

The Conocephalinae, or meadow katydids, bring me to the meadows. This subfamily consists of three North American genera, the Conocephalus (smaller meadow katydids), Orchelimum (larger meadow katydids), and Odontoxiphidium (wingless meadow katydids). The last is represented by a single species in the Southeast. The other two, with nearly forty species, are better represented. While the Conocephalus and Orchelimum are different in size, both physically resemble the monocots that surround them. Their ticks and buzzes surround *you* in the open meadows, fields, and fens, creating a soft static in the air.

Most of the insects in this group are a mixture of browns and greens and can be very difficult to see when resting on a stalk of grass. This especially goes for the larger meadow katydids, which, upon seeing you, shimmy to the opposite side of the stem. The smaller meadow katydids hold a bit less confidence in their ability blend with the background, and at your advance will often hop and fly through their grassy domain. Those are just the ones I notice more easily, though. Many do freeze in place and hide behind a stalk, as evidenced when I actually stop and look a little harder.

The females have long, swordlike ovipositors, used for slicing into the plants to lay their eggs. The female Straight-lanced Meadow Katydid *(Conocephalus strictus)* bears the mother of all ovipositors. This species is found in upland meadows throughout most of the country (absent in the far west). With an ovipositor that can be nearly double the insect's length, she appears more sword than bug. If you get too close, she will tumble to the ground and hide, head down, sword up, in the weeds.

The Coneheads *Not* from France

The head-down, tail-up pose is a defense tactic also employed by the coneheads. These insects share the open habitats with meadow katydids and can be told from them by the prominent cones emanating from the tops of their heads. There is one species, Broad-tipped Conehead *(Neoconocephalus triops),* that ventures into the treetops.

When a conehead senses danger, it drops to the base of the plant and buries its head in the tangle of blades. The grass-shaped wings make the insects indistinguishable from the surrounding blades. Many species have two color morphs—green and brown. This affords some the advantage of blending with living plants, while others meld with dead plants and darker seedheads.

One of my favorite "conehead excursions" was to the Okefenokee Swamp National Wildlife Refuge in Georgia. Having obtained the proper permits to go in at night, I set out with the words of one of the rangers ringing in my ears. I had asked if there were any snakes in the area I should be watching out for. He rattled off a couple of names and then said, "But you don't have to worry about them as long as you stay out of the tall grass." I thought to myself, "Oh good, because that's exactly where I'm going to be."

I didn't encounter any snakes (which was actually a little disappointing), but I did get to meet a couple of alligators. I was recording the call of an unknown conehead at the edge of a pond when I heard a guttural sound at my feet. I shined my flashlight to the grass, and lo and behold, there was a baby alligator! It was about a foot long and made no attempt to crawl away. Then I heard a loud splash a couple feet in front of me. It was mama gator, charging to the aid of her offspring. I then learned something about myself. I have an incredible backward long jump! I was back on the trail in less time than it took to think about it. The mother, satisfied I was far enough away, went back into the water, followed by her hatchling. I moved a little farther up the trail and found more of those mysterious coneheads. The call was a steady, shushing train of pulses, given from stunted wings. They were stocky-looking creatures, some pale green, others the color of Silly Putty. Adults and nymphs were mixed in their little colony. All had a hooked cone on the head, with a thornlike tip. They looked like nature's can openers. I'd never seen anything like it! The antennae had a striking pattern, the segments alternating in black and white, like something out of a Tim Burton movie. I didn't want to remove any animals from the sanctuary, so I came back the next day and was able to relocate the group. They were Davis's Coneheads, and my first encounter with the Belocephalus, or short-winged conehead, genus. In my mind, they will always be Alligator Coneheads (although I kind of like the idea of calling them can-opener coneheads).

There are three more Copiphorinae genera in North America, two with a single species represented, and one, Neoconocephalus, with fifteen. The latter is the group with which I am most familiar, and I consider them high among my favorites. I attribute that to their faces. Many have raspberry red "lips" accented with yellows and oranges. That, along with the cone protruding from the top of the head, gives them the appearance of little clowns, or overly made-up hussies.

One of the most common species in the Northeast is the Eastern Swordbearer *(Neoconocephalus ensiger)*. Its name makes reference to the long, swordlike ovipositor of the females. I'm not sure why this species was singled out with that moniker, since all of the coneheads have long, swordlike ovipositors. They remind me of those little plastic swords used for spearing cherries in a cocktail. The Eastern Swordbearer call is a series of soft, rapid buzzes, mimicking the rhythm of an old locomotive. They are the first coneheads I hear in summer. This is because many call from the tall grasses along the sides of the road, and I hear them while driving with the windows down. Their call puts me right in the season of summer. The temperature has become warm enough to be driving with the windows down, and the season has advanced far enough for the final molt of the Eastern Swordbearers—the timing is perfect! Their calls are often intermingled with that of the Round-tipped Conehead *(N. retusus)*. These smallest members of the group are equally common and call with a soft, continuous "live wire" buzz. As you drive down the country roads on a warm summer night, the buzzing of the round-tips follow you. You leave the range of one, while entering the range of another, creating the impression that you are not driving past them, but they are with you for the ride.

The call of the Robust Conehead *(N. robustus)* is more of an acquired taste. It consists of a loud, uniform buzz. The late orthopterist Samuel Scudder compared it to the droning of a bagpipe. I like the sound of bagpipes, but not necessarily one stuck on a single note. I don't dislike the call of the Robust Conehead, either—from a distance. They are found in the tall reeds along the edges of wetlands.

False Katydids, the Welkin Ringers!

Driving with the windows down is a great way to sample the insect calls of an area. I am admittedly a distracted driver on summer nights, prone to hit the brakes at the wisp of a new sound finding my ears. Sometimes, though, these insects will reward you with more than their song. I was driving through North Carolina one night after a very bad day. I had traveled south to hunt for Ensifera that could not be found in my more northern haunts. There are just those days when things don't seem to go your way. I had gotten a speeding ticket earlier (for going TEN miles over the speed limit!). This was after being lost for about three hours, getting stuck in horrendous traffic, and arriving too late to wangle my way into one of the state parks, where I'd hope to spend the night. I ended up in the middle of cotton country and, after hours of driving, couldn't find a place to stay. It was around midnight. I was tired and cranky, and would have killed for a cold beer. The car windows were down, of course, should I hear something calling to rescue this lost day. I stopped at a stop sign and something flew into the car and landed on top of the steering wheel. It was a Fork-tailed Bush Katydid! It just stood there, shifting back and forth on its legs, waving its long filamentous antennae. I should mention at this point that this species of katydid has always held a special place in my heart. It was the first I had come to know upon setting out to learn more about these insects. These little, gentle, unassuming leaves-with-legs are ever present on a Connecticut evening, reminding me of this with their occasional "tsip" from the surrounding shrubs. They have always done for my eyes what a gentle tune does for my ears.

As I stared at this insect on my steering wheel, I came to the realization that I was smiling. Have you ever *caught* yourself smiling? You'll be doing something, or nothing, and then notice that while you were doing, or not doing, that thing, you had a smile on your face. Noticing you are happy is even better than being happy. That's what this katydid did for me. The stress of the day was gone in the few seconds it took me to realize what I was seeing.

I looked at the insect and said, "Thanks!" I then coaxed it onto my hand and held it out the window. The katydid flew off into the shrubs at the side of the road.

The bush katydids are named for their preference of habitat. Most of them inhabit the lower herbaceous and shrubby vegetation surrounded by deciduous woodlands. There are only eight species, their genus being Scudderia. The name honors that "Father of Orthopterists," Samuel Scudder, whom you will read about later in the book. They are in the subfamily Phaneropterinae, or False Katydids, which contains just over sixty-five North American species. Phaneropterinae comes from *Phaneroptera,* which means "visible wing." This refers to the inner wings, which poke out a little past the tegmina. Of this subfamily, entomologist Willis Stanley Blatchley wrote in the early twentieth century:

> Their love calls, or songs, make the welkin ring at night from mid-July until after heavy frost.[1]

With some exceptions, like the Dichopetala, or short-winged katydids of Texas, the outer wings of the False Katydids are shaped like a leaf. The tegmina are laterally flattened. They tend to be green, but there exists a variety of color morphs from yellow to pink. The tegmina venation will also echo that of a leaf, and some species, like those in the round-headed katydid genera, can have mottling suggesting a leaf that has been long on the branch.

When compared with the ovipositors of meadow katydids and coneheads, those of the False Katydids are subtler in appearance. They are laterally flattened, broad, upwardly curved, and somewhat blunt at the end—kind of like a round-tipped linoleum knife. They are designed for depositing the flat, pumpkinseed-like eggs on branches and in leaves. Those eggs will be either glued to the substrate or slipped between the layers of a leaf.

These insects are primarily flower, pollen, and leaf eaters and are rarely found far from their food. Most call at night, but some get a head start in the late afternoon.

I am fortunate to have in my own backyard all three Connecticut False Katydids in the round-headed katydid, or Amblycorypha, genus. They inhabit the little meadow I created for them, paying their rent with song from July through November. I had always known that two of them were out there: Rattler Round-winged Katydid *(Amblycorypha rotundifolia)* and Oblong-winged Katydid *(A. oblongifolia)*. Both are common in the entire Northeast region. The third, Carinate Katydid *(A. carinata),* was always there, too. I just didn't know it. I knew Carinate Katydids existed, and that they likely existed in the area. They are nearly identical to the oblong-wings, most easily told from them by the sharper, or *carinate,* ridges along the dorsolateral edges of the pronotum. It's not exactly a feature that screams "Notice this!" and I'd searched for them for years. I came home late from a meeting one night in mid October, and espied an Oblong-winged Katydid on the side of my house. The oblong-wings had renewed their calling after a week or two of silence. It was as if they had just gotten their second wind. This individual was attracted to the porch light. When I looked at it more closely, I noticed it had those keeled ridges on the pronotum. After examining some of the other features, I concluded that I had finally found my Carinate Katydid. Then a switch in my head flicked on. The calls of the oblong-wings I was hearing sounded a little different from what I normally heard. Were they *all* Carinate Katydids? With unbridled anticipation, I grabbed my net and flashlight and captured four individuals—two males, two females—three were in a low cherry tree next to my deck. They were all Carinates! The Oblong-winged Katydids were on the wane, but were replaced by this species I'd been hunting for so long. There were still a couple of female oblong-wings out there, but the yard now belonged to their carinate cousins.

I put a lot of work into shaping my five-acre yard as wildlife habitat. I plant native grasses and shrubs, maintain a meadow by mowing only once a year, and allow for as much a mix of habitat as the area can sustain. It's nice when it pays off. I've recently discovered within my little microcosm what I *know* is a new addition—Greater Angle-wings. I know this because I've been listening for them for over ten years. These are the true beauties of the False Katydids, their wings forming a leafy green spearhead. "Rhombifolium" in their scientific name, *Microcentrum rhombifolium,* makes reference to those rhomboid tegmina. The genus *Microcentrum* is Greek for "small point." That small point is found at the tip of the hind wings, which protrude from those angular tegmina and put a sharpened point on the spearhead. There are only six Microcentrum in North America, and two in the Northeast. Their ticking call

Carinate Katydid (*Amblycorypha carinata*),
showing sharp (carinate) ridges on pronotum.

comes from the very tops of the deciduous trees. With little reason to leave their elevated perches, they are rarely seen. The exception occurs in habitats where the tallest trees are not very tall. For that reason, fruit orchards are great places to seek them out. Not only are the insects at a more accessible height, but they enjoy the fruit the trees produce. Having heard Greater Angle-wings throughout the Northeast, I expected to have them in my own yard. However, summer after summer would pass with nary a tick from the treetops. And then, just last year, I heard them; not just from one tree, but three. The Angle-wings have moved in. Why? I don't know. They do tend to congregate in more established neighborhoods. Maybe, now that mine is approaching thirty years, the time was ripe. I'm glad someone was here to notice them.

Fake Leaves, but True Katydids

The Greater Angle-wings do have company way up in those trees. The Common True Katydid is one of the four North American species in the Tettigoniidae subfamily of Pseudophyllinae (meaning "fake leaves"). They are spread out across nearly the entire eastern half of the United States. Two of those species are found only in Texas. The other one is limited to Florida and far southeast Georgia. The true katy-did most people are familiar with is *Pterophylla camellifolia*. Its croaking, grinding call forms the dominant sound wherever their arboreal turf lies. Even though, like the angle-wings, they inhabit the upper stories of deciduous trees, they are easily told apart from them by their bulging, convex tegmina. Their hind wings are kept hidden beneath those leathery tegmina and are used more for easing an emergency landing than for flying. Those landings can be the result of a heavy storm or an escape from a predator. These scenarios provide the best chance of finding one, as they may be found slowly working their way from the understory back to their lofty perches.

Ah, to Tame the Mighty Shieldback

The last members of the native katydid suborder are found as low as their angle-winged and true katydid counterparts are found high. The Tettigoniinae, or shield-backed katydids, are dwellers of the ground and low plants. Because of this, their bodies forgo the green shades of the living plants for the browns, ochres, and grays of the dead leaves across which they wander.

These are some tough-looking bugs. There are over 120 species found throughout North America. What gives them that demanding presence is the rounded, buffalo-back pronotum. The pronotum is extra long and covers the first couple basal segments of the abdomen. I suppose that humped shape, which hangs the head dramatically below the thorax, gives it the appearance of an animal about to charge or ram into something. Sometimes the most imposing creatures are actually quite gentle. That's not the case with the shield-backs. Many of them can give a strong bite, and they are known to eat weakened members of their own species.

The first shield-back encounter gave me a run for my money. I was hiking along a trail at dusk when a new call switched on. It was a lazy series of high, sputtering buzzes coming from the ground along the edges of the path. I got on my hands and knees and searched the area. The caller stopped stridulating, which meant I was close. All I managed to find was a brown grasshopper, and while I really wasn't that familiar with grasshoppers, I knew that what I had heard came from a cricket or katydid. As I continued down the trail, I heard more and more of these insects, but was unable to find them. Soon it was too dark to try.

I returned the next day at dusk, this time with a flashlight, determined to locate the mysterious singer. The calls began anew, and I crawled on the ground, checking the leaf litter and the lower branches. Again, I came across one of those brown grasshoppers. This was too much of a coincidence. I looked more closely at it this time and noticed the long antennae. How had I missed that? This was not a grasshopper, but a katydid— and a katydid that lived on the ground! I brought it home, where it continued to sing in my studio. It turned out to be a Protean Shieldback *(Atlanticus testaceus)*. I was surprised at how well it called with such dinky little wings! Since then, I've found American Shieldbacks *(A. americanus),* which have even smaller wings and yet manage to crank out an equally audible call.

In 1893 the naturalist William Thompson Davis wrote an account of a Protean Shieldback he had befriended in Staten Island, New York:

> On June 26 I heard in a moist pasture a stridulation that was unknown to me . . . In due time I discovered, in a tussock of rank swampgrass, the brown songster . . . perched on a dead leaf. He was transferred from the tussock to a tin can and at home I made a home for him in a larger can in which was a branch of post oak whose leaves soon dried, furnished innumerable nooks and crannies in which to hide. Usually, however, the insect did not hide at all but perched himself on one of the topmost leaves and there waved his antennae after the manner of all long-horned Orthoptera. Starting with raspberries, he had the rest

of the fruits in their season, including watermelon, of
which he showed a marked appreciation. If I offered
him a raspberry and then gradually drew it away, he
would follow in the direction of the departing fruit,
and would finally eat it from my hand. At night he
stridulated with unabated zeal to the first of August
when he began to be less sprightly. Finally his song,
instead of filling the room, was but a faint sound and
the end came on the tenth or eleventh of September.[2]

Ah, to tame the mighty shieldback . . .

five

The Crickets

Let us now shift from the katydids to the crickets. There are two North American families of singing crickets, the Gryllotalpidae (Mole Crickets) and the Gryllidae (True Crickets).

Family Gryllotalpidae (Mole Crickets)

The Gryllotalpidae comprise a small group, with only eight species found in the United States (one of those in Hawaii). Two of those species, Northern Mole Cricket *(Neocurtilla hexadactyla)* and Prairie Mole Cricket *(Gryllotalpa major),* are native to this area. One is historical. The rest likely arrived as stowaways in the soil of nursery stock or in the dumped ballast of ships.

If you translate "Gryllotalpidae" into English, you get "cricket mole." *Talpa* is the genus of the European Mole *(Talpa europaea).* The insect and the mammal share many traits. They are large-clawed tunnelers with velvety fur that spend most of their life beneath the ground. While the European Mole feeds almost exclusively on earthworms, mole crickets supplement their worms with plant matter and various arthropods. I mentioned the velvety fur on the cricket. Fur is a mammalian trait. The insects' fur is actually a downy pubescence and can be found in varying degrees on a number of the True Crickets as well.

The mole crickets are armed with two specialized front legs. The tarsus and tibia join to form a large claw called the *dactyl*. These digging tools are flat and heavily toothed, and, in some species, can get a stranded mole cricket beneath the ground in seconds. I witnessed this firsthand while hunting for Northern Mole Crickets. After filling a container with the sandy soil from a known mole cricket subway network, I picked one out and set it aside on the ground for a few seconds while I continued to sift through the container. Just a few moments later, I looked down and it was gone, disappeared down the hole it had just dug. I had to dig the little critter out again.

Mole cricket bodies are streamlined and dorsally flattened, designed for slipping through the soil. The antennae, unlike

Dactyl (digging claw) of Northern Mole Cricket (*Neocurtilla hexadactyla*).

those in all the other Ensifera, are short. Long antennae would not hold up to the abuse of being pushed through those tight quarters. To the human eye, males and females are virtually indistinguishable. The only way to tell them apart is by the different venation in the wings. The female's ovipositor is either vestigial or absent, so that's of no help in sorting them out. Some species of mole crickets can fly and are attracted to lights.

The male calls from his burrow, inspiring the female to join him in his lair. Sometimes she'll tunnel to him. Some fly to the opening and join him in his mating chamber. Many species lay their eggs in a special nursery chamber, which is then sealed. Some, like the Northern Mole Cricket, lay the eggs in a chamber off the main burrow, and may stay to attend them. Unlike most of the Ensifera, which overwinter as eggs, mole crickets pass this season as adults or nymphs.

While subterranean insects may seem impossible to locate, they provide the searcher with a visible clue to their presence. When mole crickets travel, the soil above them is pushed up to form a raised ridge. This is yet another similarity to moles. You will often find exit holes at the end of those ridges. The crickets come to the surface at night to feed on low plants, fruit, nuts, and decaying organic matter.

Gryllidae (Crickets)

Our most musical Ensifera are members of the family Gryllidae, or Crickets. There are eight North American subfamilies within this group, comprising about 120 species.

When most people think of crickets, they think of little, dark, ground-inhabiting insects. Although a good number of them certainly fit that bill, there are many arboreal species exhibiting a wide range of leafy colors.

It's difficult to come up with a general physical description of crickets. Some are dark, barrel-shaped pretzel nuggets, while others resemble pale, delicate stems and buds. What makes a cricket a cricket, and not, say, a katydid, is the handful of features described for the two groups in the beginning section of the Tettigoniidae in Chapter 4.

The calls of katydids tend to be drier, and lack the pitch of a cricket call. The cricket typically lifts its tegmina from the body while producing that call. Katydid tegmina stay close to the body. It is not unusual to hear crickets calling during the day. Many that do, continue into the night. Some call only at night.

The male cricket has a several different calls. One is used to call in a female. It is usually sustained, vigorous, and repetitive. Another call can be employed as a repellent to other males with the audacity to wander onto their turf. Then there is the courtship song, given in close proximity to the female. This tends to be quieter, and more intimate. Upon choosing a male with which to mate, the female approaches from behind and receives one or several small spermatophores, which are attached to the tip of her abdomen. They forgo the spermatophylax treat delivered by the katydids. Instead, many provide a different form of nuptial offering. The male ground cricket allows the female to chew his tibial spurs. These are sharp, movable spikes lining the hind legs. She nibbles at the tip, which basically opens the cap for the release of fluid. The Restless Bush Cricket *(Hapithus agitator)* offers his wings for a meal. Male tree crickets have a metanotal gland on the dorsal surface of their thorax. The gland secretes a snack into a little pit.

Tree crickets deposit their eggs in various parts of plants. The ground dwellers oviposit in the soil or rotting logs.

Most crickets overwinter in the egg stage and hatch the following spring. The nymphs inhabit the same environs they will populate as adults. As with most of the Ensifera, they tend to live among their food sources. Because crickets are so varied in habitat choice, so too is their food. They are generally omnivorous, but most make the various parts and products of plants the main course. A number of them feed on other insects—alive and dead—and other decomposing biota.

The Ubiquitous Field Crickets

Our most familiar crickets are members in the subfamily Gryllinae. These are the field crickets. They're fairly large, fairly common, and vigorous singers. They live beneath things— under logs, flat stones, and dead leaves. One of them is lucky enough to have its home on the beach. The Sand Field Cricket *(Gryllus firmus),* with a name that seems to contradict itself, is a relatively new find for me. I was birding late one afternoon behind the Dock and Dine restaurant in Old Saybrook, Connecticut. The town owns a little chunk of land there, on what is called Saybrook Point, and keeps it open the public. The month was October, and my Ensifera hunting season was beginning to ebb. As I walked a little farther up the beach, I heard crickets calling. And they were in the sand! They sounded like Fall Field Crickets, and that's what I expected to find when I rolled over a sun-bleached log. They quickly scurried to cover, but I chased a few down. I noticed right away that these crickets had sandy-colored wings. Fall Field Crickets have darker wings. The females had very long ovipositors, too, longer than what I would expect on a Fall Field Cricket. I took some home and did a little research. I was able to identify them as *Gryllus firmus,* but I knew this was primarily a southern species. It turned out that there was a record of a population of these southern species in Connecticut. That record was for Saybrook Point! I have since found them at several other beaches in my state. They're actually quite common.

Every year I look forward to seeing those tan-vested beach bums. I bring a few home, put a little sand on the bottom of the terrarium, and complete the scene with one of those little paper drink umbrellas. They provide the ukulele music.

Scaly Crickets

The Sand Field Crickets are not the only species I've been fortunate to find at the northern end of their range. In 2006, on one of my southern Ensifera hunting forays, I had pulled up at a dead end by a beach in Ocean County, New Jersey. It was dusk, and I waited in the car for the biting flies to call it a day. When darkness set in, I left the car and began to search the phragmites along the edge of the road. Seeing nothing along the outer fringes, I pushed myself in among them. It had rained earlier, and I was soon soaked to the skin, but it was warm out, so it didn't matter. I don't know why, but I had a good feeling about this habitat. That happens sometimes. It sets me in a state of mind where I don't think about wet clothes, muddy feet, sucking mosquitoes, or any of the other discomforts distracting me from my quarry.

My hope was to find some kind of conehead that couldn't be found in my home area. There were several possibilities. I found a number of Long-beaked Coneheads *(Neoconocephalus exiliscanorus),* but those live practically next door to me in Connecticut. No new coneheads were forthcoming, but I did locate a couple of Restless Bush Crickets, which I will talk about shortly. I scooped them up and was about to move to a new location when I saw a curious-looking insect on one of the reeds. It was a slender, alert little creature I had never seen before. I wasn't even sure what group of insects it belonged to. I could see that it was an Orthoptera, and a cricket of some sort, but that's about as far as I could take it. The body was long and dorsally flattened, and two long cerci—about as long as the body—extended from the tip of the abdomen. It looked a lot like a pale earwig or a silverfish. Naturally I scooped it up and brought it back to my room to key out. It turned out to be a Forest Scaly Cricket *(Cycloptilum trigonipalpum)*.

Scaly crickets are in the Gryllidae subfamily of Mogoplistinae. There are twenty species in this group, none on the East Coast found north of New Jersey. At least none have been found yet. There are a couple of central U.S. species found at a higher latitudes. The name "scaly" makes reference to the powdery scales covering the body. The females lack wings, and the males have only the tegmina, which are greatly reduced. My little Forest Scaly Cricket was a male. The maroon-tinged tegmina peeked out, barely, from beneath its flat, overlarge, teardrop-shaped pronotum. I was amazed to hear it call, which it did with gusto in my hotel room that night. It didn't seem to have enough wing to produce a song, let alone one that carried so well! The song is a series of high, buzzy "zeeps." It strikes a pitch that hits my ear just right—not too high, not too low. Listening to this "zeeping" insect in my room, I'd come to the conclusion that if I found nothing else for the rest of the week, the trip was a success.

I have since heard other scaly crickets farther south, calling from shrubs and trees, but have been unable to capture one. It is all too often the case that, in the hunt, luck plays a vital role.

Bush Crickets, Oh So Dapper

Luck certainly played a role in the next cricket found pushing the northern boundaries of its range. This one was discovered on a biodiversity "blitz," or bioblitz, sponsored by Menunkatuck Audubon of Connecticut. The purpose of these events is to set out and record every species of living thing in a given area within a twenty-four-hour period. This is designed to give a snapshot of the diversity of that area. On September 12, 2002, scores of naturalists, botanists, entomologists, mycologists, ichthyologists, and "ologists" of every other ilk set out at noon from Killam's Point in Branford. As in previous years in other towns, I teamed up with a group of naturalist friends; we call ourselves the "Corps of Discovery." While we had a productive day, and as an entire group recorded 1,722 species of all manner of biota, what awaited me back at headquarters at the end of the day was without a doubt my personal highlight of the event. Sitting in a jar on a table was a cricket one of the naturalists (Chris Sullivan) had found while surveying the insects along a stream. Because I was the "ortho" guy, it was left for me to identify. It was the most dapper cricket I'd ever seen! Its boxy body was reddish-brown, with a somewhat translucent orange coloring at the bases of the femora. It had long hind legs, which seemed as if they belonged on a larger cricket. What stood out on this insect, though, were the buttery yellow "racing stripes" along the outer edges of the tegmina. I had no idea what it was. To a naturalist, there is no greater thrill than seeing something you can't identify.

I did see it was a male, which is usually a bonus in such finds, because of the chance of hearing it sing. This male never sang, though. I brought it outdoors to photograph, and it sat on the leaf, with apparent patience, while I took my fill of pictures. I had no idea that in the time between setting it on the leaf, and the five minutes of shooting, it had died! I have no idea why, but felt fortunate that I was able to get to it while I did.

The species turned out to be a Restless Bush Cricket *(Hapithus agitator),* and it was the first one found in Connecticut. I've since found males and females in the southern states, and another was recently found in Connecticut. They are in the Gryllidae subfamily of Eneopterinae, which, with only eleven species in North America, is a fairly small group. All but two species are limited to the Southeast. The other, Jumping Bush Cricket *(Orocharis saltator),* is absent north of Connecticut, but its range extends to the central United States.

The Eneopterinae, or Bush Crickets, inhabit shrubs, herbaceous plants, and low, deciduous trees. They feed upon the flowers, leaves, and fruit of these plants. I was familiar with the call of the Jumping Bush Cricket; its rich chirp is heard in city trees and country shrubs throughout half the country. The reason its cousin, the Restless Bush Cricket, never called for me was because they do not call! They have tegmina, but the northernmost populations don't use them for song. They are instead offered as a meal for their mates. The female mounts the male from the rear, and while he is slipping a spermatophore to her venter from underneath, she is chewing on his wings. This keeps her too distracted to remove and eat the spermatophore, which she is otherwise likely to do, and holds her attention long enough for the insemination to take place. Because the male crickets actively hunt out their mates, as opposed to calling them in, the use of their tegmina can be relegated to this purpose. A male hunting on foot for females is the reverse of how the Orthoptera typically search for a mate. It probably helps that their colonies tend to stay in one area. In 1962, orthopterist Richard Alexander wrote, "I know of a colony of *Hapithus agitator* only a few yards in diameter which has not moved in seven years. When a male of these crickets locates a female, he stays with her for hours, or even days—following her about wherever she goes and courting whenever she stops for a moment."[1]

The full fresh, unchewed wings on the male cricket found on that bioblitz is sad proof that it died a virgin.

Ground Crickets, "Dwellers of the Glade"

As mentioned earlier, male ground crickets also give of them-selves, literally, when mating with females. Ground crickets are in the rather large subfamily Nemobiinae. The scientific name is of Greek origin and means "forest or glade dweller." However, because many are found in or along the edges of forests, they may have been better called "Edafosiinae," or something of the sort, meaning "ground dweller." These tiny crickets, looking at first glance like miniature field crickets, live in the soil and sand of field and forest. One species, the Sphagnum Ground Cricket *(Neonemobius palustris),* inhabits open sphagnum bogs. Another, the Gray Ground Cricket *(Allonemobius griseus),* prefers dry, sparsely vegetated sandy habitats. Tinkling Ground Crickets *(A. tinnulus)* "tink" away in sun-swept leaf litter along the edges of woods. If there is an ecological niche to be filled, there is likely a ground cricket up for the job.

Prior to mating, the male calls to attract females. Once a female is drawn to the caller, she moves in a little closer. The male's call changes, and he rocks back and forth on his feet, a further act of enticement. He then turns and backs into the female, who, if impressed by why she sees and hears, climbs onto the male's back. He stops calling and lifts one of his hind legs. It is held in a position to keep the female on his back while he inseminates her from beneath. The female bites the tip off of the highest spur on his tibia. This releases a secretion, upon which she feeds. The key to this working is for the male to hold his "knee" high enough to force the female to climb far enough up his back for their two venters to come into contact. This secretion no doubt has some compelling quali-ties to keep her place. In fact, other females will sometimes take a nip at that elixir-giving spur, even though they have no intention of mating with the male. The male, in this case, is not willing to allow her to partake. Yes, human analogies come to mind.

The fertilized eggs are laid in the ground and hatch the following spring.

In Chapter 8, I write about the ethereal song of the Sphagnum Ground Cricket. That species takes a bit of work to find, although it is well worth the effort. I have two other species in my yard, though, that offer an equally sweet song. To hear them, I just need to open my window. Allard's *(A. allardi)* and Tinkling Ground Crickets are stationed around the edges of the woods in my backyard. The former sings with a high, sustained trill, broken by brief pauses. The latter "tinks" like a tiny glass bell. They both sing in a similar pitch, but they can be told apart by the velocity of the stridulation. You can hear each individual note of the Tinkling Ground Cricket, and, should you be so inclined, could count them. The Allard's notes run together. Of course, this is in the ideal situation. A slow calling Allard's or a fast-calling Tinkling could be confused with one another. This is only frustrating if you care. Otherwise, these two closely related species could be enjoyed as "just one of those two." They are fairly easy to distinguish by sight. The Tinkling Ground Crickets have a clean, orangey head, unlike the dark brown, lightly striped heads of the Allard's.

The Spidery Trigs

There is a little cricket in my area very similar in appearance to a Nemobiinae, but it's placed within a different subfamily. The first time I came across one was while net sweeping at Hurd State Park along the Connecticut River. Upon checking my net to see what I caught, I noticed a little cricket crawling up the side. It looked different from some of the other ground crickets

I'd seen, so I put it in a glass jar. I held the jar to my face for a closer look, and was astonished to see the cricket crawling, like a spider, up the side of the glass! Ground crickets don't do that. They can crawl up things, but not smooth, vertical surfaces. This immediately told me I'd come across something new.

My climbing cricket turned out to be a Say's Trig *(Anaxipha exigua)*, a member of the Gryllidae subfamily of Trigonidiinae. It is named after the notable nineteenth-century naturalist Thomas Say. In researching this insect, I'd come across a paper written by entomologist Bentley Fulton in 1956.[2] In the introduction he observed, "Unlike true ground crickets, such as *Nemobius,* they are able to walk up a glass jar." I remember saying aloud, "Oh come on!" Not only had I found a description of my trig, but a unique shared observation—the very one that led me to further research this insect.

The Trigonidiinae, or trigs, or sword-tailed crickets, or winged bush crickets—they have a few common names—are a relatively small group with four genera in the United States. They tend to prefer humid habitats, where they live on the foliage and stems of tall grasses, herbaceous plants, shrubs, and trees. Some can skim along the surface film on still water. They are spry little insects with bulging eyes and a range of colors. The females have a saberlike ovipositor, which is hinted at in the Greek origins of two of their genera names, Anaxipha ("upraised sword") and Cyrtoxipha ("curved sword"). Those two groups are told apart by, among other things, their color. The Anaxipha are brown, and the Cyrtoxipha are green. One species, *Phyllopalpus pulchellus,* or Handsome Trig, is multicolored. Its head and thorax are a rich, deep cherry red, and the body and tegmina are blue-black. If you are fortunate to come across one of these, have a look at the antennae. They are black at the base, white in the center, and then black again at the tips. Also look closely at the palps, the ends of which are enlarged

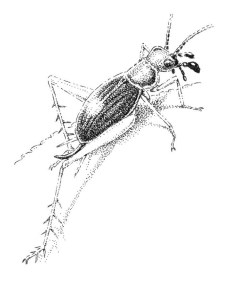

Handsome Trig *(Phyllopalpus pulchellus).*

like paddles, or leaves. I read one description that compared them to boxing gloves. The genus *Phyllopalpus* means "leaf-shaped mouth." If approached cautiously, these trigs can be seen rapidly vibrating those palps on the substrate.

Upon first glance, these insects look like beetles. This is due to the convex tegmina, a unique feature in North American Ensifera. The only others to share that feature so dramatically are the true katydids. Their tegmina bulge out, too, but from the sides.

The call of this cricket is a stuttery trill and has been compared to that of the Four-spotted Tree Cricket.

Tree Crickets, the Singing Stems

This brings us to our last subfamily, the Oecanthinae, or tree crickets.

The songs of these slender little insects play a large role in filling in the background sounds of a summer and early autumn evening. We so rarely get to see the singers, though. They are well camouflaged, in their pale greens and browns, within the foliage of the flora. To my eye they resemble delicate plant stems or buds. The body is flattened dorsally, and the head is held horizontally. The crickets in the other subfamilies hold their heads vertically. This horizontal position allows the insect to hug the leaf more tightly, thereby reducing its profile and making it harder to spot by a passing predator.

While the common name for these insects is "tree crickets," most of the twenty North American species can be found in shrubs and low, herbaceous plants. Many call during the day, making them easier to locate. There are some species, though, like the Two-spotted Tree Cricket *(Neoxabea bipunctata),* that spend most of their lives in the upper canopy.

The male Oecanthinae have wide, translucent wings, which rest flat on the body. Most of the wing area serves as the stridulary organ. When they call, those wings are held at about a 45-degree angle from the body, forming, most notably in the wider-winged species, the shape of a heart. They will often chew a hole through the middle of a leaf, poke their head and wings through to the top, and call through that hole.

The females are more slender, and their wings wrap around their body, which gives them a somewhat elongated diamond shape.

The tree cricket mating ritual bears some resemblance to that of *Hapithus agitator* and the ground crickets. The male lifts his tegmina and allows the female to climb on his back from behind. This exposes the dorsal section of his thorax known as the *metanotum.* There is a little pit in this area, which is fed a secretion from a specialized gland called the *metanotal gland.* The female can't get enough of that stuff, and while she is sipping away, he slips his spermatophore from the tip of his abdomen to the genital opening of the female. The first thing the female will do upon uncoupling is remove and eat that spermatophore. The offer of that nuptial treat could entice her for another try or two, or keep her occupied long enough for enough of the sperm to enter and inseminate her.

The female ovipositor is needle-shaped, designed for drilling. Prior to laying her eggs, the female chews a little depression in the substrate. This could be bark, a stem, or a pithy twig. She then pushes her ovipositor in that chewed area, bores a hole, and deposits the eggs inside. Once the eggs are laid, the hole is filled with a mucilaginous substance. That hole is then plugged up with chewed plant material, or her excrement.

Many tree crickets call in the middle of the day. I remember coming across a daytime gathering of Fast-calling Tree Crickets *(Oecanthus celerinictus)* in the low weeds outside of Fort

Female (only her antennae are visible) Black-horned Tree Cricket
(Oecanthus nigricornis) feeding from metanotal gland of male.

Pulaski in Savannah, Georgia. Most observers fail to notice the trait from which this cricket's name is gleaned. They actually sounded a lot like the closely related Black-horned *(O. nigricornis)* and Four-spotted Tree Crickets *(O. quadripunctatus)*. If one of those were calling from a nearby plant, then the difference could be heard. "Celerinictus" contains the Latin root word *celer,* meaning "swift." Just think of the word *accelerate.* The trill of this species contains more pulses per second than others in this subfamily.

I thought I did sense a rapidity of the calls I was hearing, but it was also about 95 degrees outside. Higher temperatures increase the speed of most Ensifera stridulation. I pulled to the side of the road and walked through the weeds to seek out the singers. Within seconds, blood was pouring out of my shins. There is some kind of vine growing along the sides of the road—never got a chance to look it up—that grabs your legs and rakes your shins with long, sharp thorns. I had absolutely nothing to mop up the blood, and it wasn't stopping on its own, so I climbed into the car to drive to a nearby convenience store for napkins. Once I had the bleeding under control, I returned to the spot where the crickets were still trilling away, showing no obvious concern about how I had made out.

I quickly located several males. They were not at all shy about singing in front of me. Females were found in the plants around the males. It turned out to be quite a hotbed of Fast-calling Tree Crickets. I captured a few in a jar and put them in the car. Later that night, as I was driving to find something to eat, I heard a different call coming from the backseat, where I had my cache of different Ensifera gathered from the previous days. The songster trilled for three seconds; then stopped for ten seconds; trilled for three seconds; stopped for ten seconds . . . It repeated this with various lengths of silence for about ten minutes. It was my *Celerinictus,* sounding very much like a higher-pitched Davis's Tree Cricket *(O. niveus)*. Every reference I have read, and heard, of this species describes their call as being an "uninterrupted" trill. And here's this male proving he'd never read any of those descriptions. I thought that maybe this was a courtship version of his call, but there were no females in his cage.

Aberrations make the world go 'round.

My favorite tree cricket story involves my late cat, Jupiter. She was getting on in years and spent a lot of time sleeping. Her favorite spot was next to a window in my studio. One summer evening, as I was working at my drawing table, she started snoring. She had never snored before, and I attributed it to her advancing age. Every night it was the same thing—I'd be working in the studio, with Jupiter on the other side, steadily snoring away. It wasn't a loud snore, and I didn't pay much attention to it because I usually had music playing.

Then one evening, I looked over at the snoring cat, and she wasn't there. The snoring, however, continued. I looked under the couch and checked the rest of the room, but the cat was elsewhere. The snoring, I realized, was coming from outside, and it wasn't so much of a snore as a trill. As it turned out, there was a buddleia bush next to the window, where a Narrow-winged Tree Cricket *(Oecanthus niveus)* had taken up residence. The call is a steady, pulsing trill, with the pauses and trilling taking up about the same amount of space. It was the perfect rhythm of an aging, slow-breathing feline.

I would like to think Jupiter had selected that spot because she liked that sound. It *was* very soothing, and she seemed to go to that spot only in the evening. However, her hearing was not that great. It really doesn't matter. In my mind, the two will be forever linked. When I hear Narrow-winged Tree Crickets now, I can't help but associate the sound with that sleeping old cat.

six

The Mighty Cricket Gladiators

Christmas Eve, 1967. Unlike the many families who opened their presents on Christmas day, the Himmelman family celebrated this holiday on the evening before. This was when Santa came. We'd actually get to see him, too. Santa would burst in through the back door bearing a big white pillowcase stuffed with presents for my brothers and me. He was always a bit thinner than he appeared on Christmas cards and cartoons, his pillow-stuffed shirt giving only a suggestion of the Cringle rotundness. His face was completely covered with a cottony beard and long white hair resembling the tousled mop atop the wizened head of Grandmamma from the Adamms Family. My father wanted to make sure he was not recognized. He never was.

On this particular Christmas, old Saint Nick brought me a most curious gift: a cricket farm. I had never heard of a cricket farm before, but assumed it would be something similar to the ant farm I had received the previous year. As with the ant farm, the "farm" came with no farmhands. You had to order them. Of course, now, to you and me, this makes perfect sense. These *toys* could spend months or years on a store's shelf. Had the ants actually been included, they'd have to do away with the word *farm* and market them as ant cemeteries.

Giving an eight-year-old gifts he could not use right away was downright cruel. At that time I was a fanatical bug watcher. I had all kinds of little critters growing in a wide assortment of bottles and jars in my room. The arrival of cold weather, however, put an end to those pursuits. Growing up on Long Island, New York, where winters kept the bugs under wraps for a few months, meant I would have to give up my hobby until April. That is, until I received that ant farm.

I remember when the ants came in the mail. I don't know what kind they were, but they were different from the ants I was used to seeing outside. They may have been Harvester Ants, which today are the ant of choice for these setups. The ants came in a long, glass vial with a cork in the open end. I brought the ant farm down into the dining room while my mother held on to the container of insects. I'm sure she felt that a seven-year-old should not be walking around the house with a glass vial filled with ants. I set the farm on the table. My mother uncorked the tube and placed the open end directly over the opening of the farm. My face was practically pressed to the glass wall of the enclosure as I anticipated the airdrop of its soon-to-be inhabitants. After four or five ants dropped into the farm, the rest of the ants became somewhat unruly. With a sudden burst of speed, they swarmed from the inside of the test tube, around the lip, and back up via the outside of the tube. They had already reached my mother's hand before she dropped the tube and sent the ants scattering across the dining room table and floor. What followed was a frantic dance of slapping and stomping, resulting in a somewhat understaffed farm.

I remedied this shortage with the addition of Carpenter Ants I wrangled outdoors. Little did I know that my once-peaceful farm would be the arena for the most horrific battle I had yet to witness. It was farmers versus carpenters. I don't remember who won, but I learned something. Ants don't always play nice.

Then came the cricket farm. This was basically a clear plastic rectangular box, similar to the containers grocery stores use to hold cherry tomatoes. Small holes peppered the lid to allow for air exchange. The kit came with rearing directions, cricket food, and a most peculiar addition—six little plastic, lime-green chariots. The instruction booklet showed illustrations of crickets pulling the chariots, and encouraged the proud owner of this product to hold cricket races. I was intrigued, but first it was necessary to order the noble steeds required to pull the carriages. I sent in the order form and waited. It took two and a half months for them to arrive. I remember that. I remember it because my mother had to call the company to see what the holdup was. They claimed they had sent it out weeks ago and promised to send another batch if they did not arrive shortly. A second phone call a week later initiated a new delivery. I can't help but wonder who got my first order of crickets, and what the recipients thought upon opening the package.

I had given up on them by the time they arrived. They came in a jar of about fifteen or so, three of them, maybe more, no longer among the living. It was hard to tell. The corpses were torn apart, evidence of the proclivity of this group to eat their fallen brethren. When I loosened the top of the jar, I was introduced to a very unpleasant smell. It was the first time I realized that insects could actually have an odor. To this day, when I see House Crickets, the species they sent, I get an olfactory *déjà vu*. It's a musty smell, mingled with the essence of rotting fruit.

I brought the crickets up to my room, where their new clear plastic home had long been waiting. Not surprisingly, my mother didn't commandeer the transfer this time, so I was on my own. I opened the top of the cricket farm, unscrewed the container filled with crickets, and dumped them in. Piece of cake! There was no doubled-back attack. Then, within about two seconds, I learned something else about crickets. They are good jumpers. Though none of them could breach the rim from the floor of their new home, some managed to do so from the backs of the others and from the little food dish. Three were liberated; two permanently. To this day, their descendants could be regaling the current homeowners with their song.

I remember well my first night with the crickets. They began chirping at dusk and were still going strong as I drifted off to sleep. Not all of them sang—just two, maybe three—which surprised me. Why weren't they all singing? I loved hearing them, though! They were very loud, but it was a comforting sound. As far as I was concerned, they were worth the wait.

The next day, it was time for the big event—the chariot race! Much as I tried, I could not picture how this would look. Would they run in circles, pulling their chariots behind them? Would they leap, their burden in tow—airborne as a sleigh lifted by reindeer? I took one of the chariots out of the package and inspected it closely. Two plastic rods extended from the main carriage. They were designed to pinch the sides of the cricket's thorax. This would provide, in theory, enough pressure to hold the chariot onto the cricket without harming the insect.

But first I needed the area in which they would race. This was accomplished with toy wooden blocks. I arranged them to form a small circular wall and even included a couple of tunnels made from those blocks with the semicircle cut from the bottom.

This completed, I then turned to hitching the chariots to my steeds. I pulled out the crickets, one at a time, until I had three of them in the arena. Then, pinning the first one down with my thumb and forefinger, I slipped on the chariot. It actually stayed attached to the cricket! However, as soon as the cricket hopped, the chariot fell off. One cricket made it about two hops before losing its burden. I tried this over and over, using different crickets with different chariots. While I became quite adept and handling these little orthopterans, my luck in turning them into potential conveyers of gladiators did not improve. At one point I tried using a little Scotch tape. This actually worked a little better, but the look of the tape tainted the experience, and they soon slipped out of that, too. It occurred to me that I had some paste in my top draw, but I nixed that idea because it would take too long to dry.

So much for the great cricket chariot race.

I wonder what possessed the makers of those cricket farms to come up with this product. There had to have been a point when someone in the company said, "Hey, I have an idea. Let's attach little chariots to crickets!" Then, I would imagine, someone had to figure out how to do it. As far as I'm concerned, they didn't do a very good job of it, and they must have known that it didn't work.

Perhaps thoughts of chariots were in the air. *Ben Hur* had come out a few years prior and was a huge hit. This naturally set off a rash of similarly themed movies and merchandise, as Hollywood is wont to produce.

It is also possible the cricket farm people knew a little something about cricket history. Crickets have been present in our culture for a very long time. Most of their association with us comes as a result of their ability to create song. There is also, however, another cricket quality that brought them along the path with humans. While the annals of cricketdom contain no records of these insects pulling gladiators in arenas, it is rife with stories of crickets as the gladiators themselves. Put two male crickets in a confined area, and you are likely to be left with one male cricket.

Male crickets have long been known to be fiercely territorial. Their call is given not only to attract females but also to

From Insect-Musicians and Cricket Champions of China, 1927,
Berthold Laufer Field Museum, Department of Anthropology, Leaflet 22.

ward off other males. Should one male ignore another's audible warning to stay off its turf, physical aggression ensues. With the field crickets and House Crickets, it often begins with feeling each other out in an antennae-fencing match. This is followed by head butting. If the intruder doesn't get the obvious message, it is punctuated with a few sharp kicks of the hind leg. Then they go at it, face to face—*grillo a grillo*—interlocking forelimbs and snapping mandibles. If the weaker cricket senses it hasn't a chance, it retreats. Should it continue to fight, it does so at risk of loss of limbs, antennae, or ultimately, its head.

Humans, also known to possess competitive and, sometimes violent, qualities, found a way to satisfy these tendencies without personal loss of life or limb. We watch others fight. Often these others are human; boxers, wrestlers, and historically, gladiators. They can also be animals, set against each other in cockfights, dogfights, fish fights, bear baiting, and other battles between a shamefully large variety and combination of species. Crickets not only made the cut to join the list of recreational combatants, but they are among the relatively few that continue to do battle openly in arenas today.

Records of cricket fighting originate in China and go back to the Tang dynasty, which spanned AD 618 to 907. Prior to that, crickets, along with katydids, were valued for their song and kept as pets. In fact, they still are. They are collected in the wild or reared from reliable stock, and kept in little cages within the home. To add to the resonance of the song, some owners will spread a small coat of wax on the wing covers.

Keeping and raising crickets for song led to a better understanding of their habits, and it undoubtedly exposed the males' penchant for fighting one another. By the Song dynasty (AD 960–1278), cricket fighting was a popular sport, attended by emperors and peasants alike. Bi Lü's seventeenth-century book *Ming Chao Xiao Shi* (The Minor History of the Ming Dynasty) explores this long history. He writes of the "Cricket Emperor," Ming Xuan-Zhong, who each year was the recipient of thousands of crickets sent by the people. These were offered as a tribute to the emperor, who would select from these gifts the best fighters. Those who made a gift of the chosen warriors could receive a good horse in exchange. Lü writes of the tragic fate of the emperor's wife: One of his prized crickets escaped when she peeked to look at it. It was then eaten by a rooster. Fearing her husband's reaction, she committed suicide. When Ming came home to see his wife dead and cricket gone, he followed suit and took his own life.

It is unclear in reading this if his suicide was a result of the death of his cricket or the death of his wife—or both—but the point is well taken: crickets were cherished. Xuan-Zhong's interest in battling crickets ignited a passion for this pastime throughout China. Here was an "emperor's sport" that could be shared by anyone, regardless of social status. Many believed that the crickets they raised were incarnations of warriors past. They were not rearing crickets, but training soldiers.

Adding to the allure of this sport is the popular legend of a Buddhist monk named Ji Gong. As the story goes, Ji Gong entered his puny little cricket in a match against a behemoth of an insect owned by a greedy, and very wealthy, landlord. It has all the makings of a David verses Goliath story, with a similar outcome. Ji Gong's cricket sent the monster packing, and the monk took his huge winnings and distributed them among the poor.

Here, the cricket became the great equalizer. The poor could not only be put on the same footing as the rich, but could best them in what was once considered their game. All you needed were some crickets, an arena, and people willing to bet on the outcome of a match. The crickets were fairly easy to obtain, although some regions were noted for producing better crickets than others. The arena often consisted of shallow clay bowl. The gamblers came from the same walks of life as they do today.

This sport also brought out the worst in people. The prospect of a quick buck, or more applicably, *Jiaozi,* took what was considered both fashionable and an art form and turned it into a seedy pastime. The gambling aspect superseded sportsmanship, and cricket fighting took on a bad name. Families went broke. Cheating was rampant, and the level of highbred crickets was diminished by people who knew nothing about them. No longer thought of as the sport of emperors, cricket fighting was decreed illegal during much of the Qing dynasty (AD 1644–1911). This was a long dynasty, though, and the sport enjoyed brief resurrections throughout those centuries.

Cricket fighting is experiencing resurgence today. It is flourishing in China's financial hub, most of the practitioners being middle-aged, unemployed men. They bet tens of thousands of Yaun on the bouts. Gambling is illegal in China, but the sport continues in side streets and other low-profile venues. Some places have little cameras set up in the arena to transmit video images of the action to monitors mounted around the room. Naturally, with gambling there is always the threat of the criminal element slipping into the mix. In December 2004, Shanghai police detained 46 individuals connected with a cricket gambling ring. In August of that same year, Hong Kong officials arrested 115 men involved in an underground street cricket syndicate. They were not arrested for the sport, but for the gambling end of it. In 2008, police arrested 66 people and seized 520,000 Yuan ($72,000) at an illegal cricket fight. Stories like this continue to appear in the news. I can see the dialogue in the prison:

"So, what are you in for?"

"Crickets."

Many different species of crickets are used in battle. A large body, a big head and jaws, strong back legs, and a shiny black face are standards of excellence. The insects are always males. Two females in the arena would get along just fine. A male and a female, *even better.* One commonly used species is *Gryllus bimaculatus,* aka the Vietnamese Fighting Cricket. In China it is called the Mirrored Cricket. In other places, it is the Mediterranean Field Cricket. These are closely related to our own field crickets, which, by the way, are the species of choice in the few places this takes place in the United States. This cricket is overall dark brown—almost black—with cream-colored wings. Another is *Velarifictorus micado,* known in China as the "Cu Zhi" Fighting Cricket. It appeared on North America's doorstep about fifty years ago and is known here as the Japanese Burrowing Cricket. It's about the same size as your average House Cricket. I write about my hunt for this species (seeking not a fighter but a singer) in Chapter 8. Many of the champion crickets come from China's Shandong Province. This 60,000-square-mile territory on the eastern border is a monsoon magnet. Thousands who make the annual trek to this region to purchase fighting crickets from the local farmers swear that its water and soil breed the best contenders. For the farmers, these sales are their main source of income.

Whatever the species, and from wherever they came, many of these combatants are as pampered as the prized athletes, human and animal, of today. They live in ornate cages of gold, china, or bamboo, or in beautifully carved gourds or boxes. Crickets that prefer a cooler environment are kept in ice-filled foam boxes. Care is taken that they do not injure themselves before a fight, to the extent that their boxes may be lined with shock-absorbing springs to prevent jostling.

They are fed only the finest of cricket cuisine, too—steamed rice, fresh apples, cucumber, seeds, corn flour, and for strength, ginseng. Calcium tablets are used to toughen the outer shell. Some practitioners allow themselves to be fed upon by mosquitoes. The blood-filled mosquitoes are then fed to the crickets. This offers a hearty meal for the insects and must provide the owner with a vision of creating some kind of blood lust in his charge. Many crickets are forced to fast prior to a bout. If they appear too sluggish, they'll be rejuvenated with a quick, energy-boosting meal of tiny insects.

Cricket-fighting season is autumn: between China's "White Dew" (early September) and "Frost's Descent" (late October). This period is when crickets are believed to have reached peak fighting potential in their four- to five-month life span.

The contenders are weighed in prior to the fight. They are then sorted into one of the three weight classes: light, middle, and heavy. Bets are placed and the two opposing crickets are shaken from their cages, or scooped with a special spoon from their gourd. Sometimes the crickets just ignore each other. One may just keep away from its enemy, knowing it doesn't stand a chance, or maybe it's just not in a fighting mood. This makes for a very boring match, so a "tickler" is used to goad the reluctant opponent into fighting. These can be as ornate as the cages and gourds, with ivory handles holding a sparse brush of three or more hairs from a rat or rabbit. Some use mouse whiskers. The simplest design employed is a piece of grass plucked from outside the venue. The idea is to urge the cricket to fight without actually pushing it into the other cricket. It takes a gentle touch, hence the gentle tool.

In the Chinese book *An Illustrated Guide to Crickets,* authored by Bian Wenhua and Yang Ping, fighting is broken down into three styles. In "Creep like a tiger, fight like a snake," the cricket slowly stalks its challenger, looking for an opening for attack. When the cricket sits still, waiting to hear the other cricket before ambushing it, it is said to be using the "Listen for sound, look for the enemy" technique. The style attributed with the greatest success is "Force of a fine steed." Here the cricket goes all out, stealth be damned, charging "like the wind" at its opponent.

The fight itself can last between a few seconds and forty-five minutes. The crickets jump, flip each other over, latch onto body parts, and sometimes, to the delight of the crowd, rise belly to belly on their hind legs as they vie for the advantage. Scientists Hans A. Hofmann and Klaus Schildberger have written about the six different stages of battle they witnessed in the cricket battle domes they constructed in Germany.[1] They are as follows: One animal retreats immediately (level 1), the contestants initially fence with their antennae (level 2) and then display spread mandibles (level 3: unilateral; level 4: mutual), which later interlock (level 5) before the animals wrestle (level 6).

If the weaker cricket does not concede (that is, run away), it is often killed or left so badly maimed that it dies. The loser is usually unceremoniously tossed away. If it was a great warrior, a "general" (a rank given to the better fighters), its loss will be mourned. Funerals for these crickets are not unheard of.

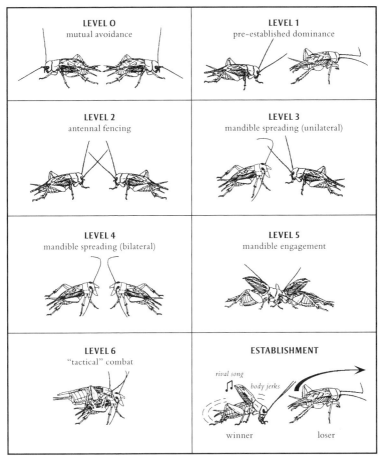

Pictogram illustrating the stereotyped escalating sequence of motor performances characteristic of aggressive encounters between male crickets.

Artist: P. A. Stevenson, from Paul A. Stevenson, Hans A. Hofmann, Korinna Schoch, and Klaus Schildberger, "The Fight and Flight Responses of Crickets Depleted of Biogenic Amines," *Journal of Neurobiology* 43 (2000): 107–120. (Used with permission.)

However, the weaker cricket often does concede. If it refuses to get back into the fight, the other wins by a TKO, but the owner of the cowering cricket has one last trick up his sleeve. He will pick up his cricket, shake it in his cupped hands, and give it a few tosses in the air. This reinfuses the fighting spirit, and it is tossed back into the arena.

This practice of introducing aggression into an insect screaming *"No mas, no mas!"* intrigued Hofmann and entomologist Paul A. Stevenson, who were spending a lot of time with crickets and this sport. They had seen how the Chinese men would shake the crickets and toss them in the air to get them back in the fight. In an article for *Nature,* they write of their initial puzzlement as to why a defeated cricket, void of all aggression for about twenty-four hours after catching the worst end of a fight, regains that fighting spirit.[2] The two men were able to replicate this in the lab and discovered that it was not so much the shaking that got them going, but the tossing. If they could get the cricket to open its wings to fly, even for a few seconds, it would reawaken the mechanisms of aggression. They concluded that something in the act of flying sends a signal from the thorax, where the flight muscles are located, to the part of the brain that controls aggression. That signal, in effect, flips the "fight" switch back on. According to the authors, "The reset of aggression by flying is the only example we know of in which activation of a specific motor pattern immediately affects an unrelated subsequent behavior." This got them wondering if this knowledge could be applied to human behavior.

This is well beyond my range of knowledge, and the scope of this book, but I do know this: there are a lot of animals out there that would be ready for a fight after a good shake and toss. A picture comes to mind of a boxer in a ring, his trainer grabbing him by the shoulders. "Get back in there," he shouts, shaking the fight back into him. It's a familiar scene. It feels like a natural action to incite aggression. Perhaps the Chinese men who discovered this were acting on instinctive knowledge. Or maybe they were just angry at the cowardly cricket.

In their frustration they shook the daylights out of it, only to learn it inspired readiness for another round.

Frustration can run high in these events. Some people make their living on the winnings of their prized fighters. A good cricket can fight in up to ten matches before becoming mortally wounded in a match, or dying of old age. Most, even the better fighters, don't make it much past five or six matches. People also make good money on selling the crickets. In the Wanshang Flower and Bird Market, there are over two hundred cricket vendors. They sell an average of $12,000 worth of crickets a day! A good fighter can go for the equivalent of a couple thousand dollars. An average cricket, though, costs about a quarter.

When I think of the crickets that arrived in the mail on that winter afternoon, aside from that smell I described earlier, I do remember noting that most of them had those needle-like ovipositors extending from the abdomen. I remember this so well because I found them to be a bit . . . unsettling. They just looked way too much like stingers, for my comfort. I knew they didn't sting, but my brain had to get past what my body was telling me. My body sensed pointy things coming out of the end of bugs and screamed "Stingers! Stingers!" I did overcome it, though, after forcing myself to hold one cupped in my hands. The more I did it, the more comfortable I became.

This could explain, however, why I heard only a couple of crickets calling from my farm at night. Those were the males. Females don't call. I don't think it's a stretch to assume that those dead crickets I saw on arrival were males. A dark battle ensued on their journey from wherever they were shipped from to that little cape in Oceanside, New York. The few surviving males had something in them that got them through it. Maybe they were better fighters. Maybe they were better fight *avoiders,* hiding behind the females. It's safe to say that there was a lot of jostling going on, *en route* to my house, activating those aggression transmitters.

The crickets' numbers dwindled over the next few weeks. I remember one night hearing just a single cricket calling. Shortly after that, there were no more calls. I released the remaining two crickets (the ones with those pointy ovipositors) into my backyard. Their little green chariots sat in their original plastic bag in my top drawer for years to follow. I had no desire to throw them out. While they never really worked the way I'd hoped, the image of crickets pulling chariots never left my head. It was such an unlikely marriage of elements, one that if pulled off successfully could have resulted in something bordering art. Had that entered the mind of the eight-year-old boy who had received a jar of crickets from Santa? No; at least not consciously. I just thought it would be a really cool thing to see.

Maybe if they made the chariots so the two poles hooked *over* the thorax . . .

seven

"Give a Little Whistle"

More Stories of the Ensifera and Us

As I child, I'd become quite familiar with crickets. I raised them in my room, and I spent endless hours on my stomach, watching them in the grass. Perhaps that's why the most famous cricket of the time looked kind of odd to me. In fact, it didn't really look like a cricket at all. For one, it was green, or sometimes tan—more like a grasshopper. I was able to forgive that it walked on hind legs, and wore clothes, but not the fact that it only had two arms and two legs. To me, it looked more like a young man with a grasshopper face.

That being said, he did have an engaging personality, and that trait stole any cartoon scene he was in. His name was Jiminy, Jiminy Cricket. I got to know him as a frequent guest on the *Wonderful World of Disney* and listened over and over to his song (sung by Cliff Edwards) "Give a Little Whistle" on a Disney album my mother had bought me.

Jiminy Cricket was a nattily dressed orthopteran in his purple top hat, tails, spats, and white gloves. He clutched an ever-present umbrella, which was used more than once as a parachute. He began as the conscience of Pinocchio in the same-named animated film of 1940. He was always on call to help the little wooden boy, who, to summon him, just had to "give a little whistle." Disney saw Jiminy as a breakout character and plugged him into their productions whenever they could. In 1998 Disney added a new cricket to the family—Cri-kee, who was given to Mulan for good luck. Now *he* looks like a cricket—"Disneyized," of course, but definitely a cricket.

The original Jiminy was based on the *Grillo Parlante,* or Talking Cricket, who appeared in Carlo Collodi's 1883 *The Adventures of Pinocchio.* This earlier version depicts the little insect as a darker, more philosophical figure. There is no "wishing upon a star" in Pinocchio's sidekick, who is actually less of a sidekick as he only appears in four of the thirty-six chapters. The Talking Cricket had lived for over one hundred years in the Geppetto home. He had little patience for the wooden boy's wild antics and said to him, "Woe to boys who refuse to obey their parents and run away from home! They will never be happy in this world, and when they are older they will be very sorry for it."

Pinocchio would have none of that, especially from a preachy little bug. The cricket goaded him just a little too far, though, telling him he felt sorry for him because he had a wooden head:

> At these last words, Pinocchio jumped up in a fury, took a hammer from the bench, and threw it with all his strength at the Talking Cricket.
>
> Perhaps he did not think he would strike it. But, sad to relate, my dear children, he did hit the Cricket, straight on its head.
>
> With a last weak "cri-cri-cri" the poor Cricket fell from the wall, dead!

What? Pinocchio killed Jiminy? That murdering bastard! Little did the marionette know, however, that the cricket would come back to haunt him as a ghost. He appears later in the book as a "tiny insect glimmering on the trunk of a tree."

"Grillo Parlante."
Artist: Enrico Mazzanti, from *Le avventure di Pinocchio*, 1883.

He warns the still obstinate Pinocchio that boys who insist on having their own way sooner or later come to grief. It is not until the last chapter that Pinocchio finally sees the wisdom in the words of Talking Cricket.

Using a cricket to personify an individual's conscience works on at least four levels: First, you have a creature that makes a sound audible to humans. You need that to be able to hear his advice. Second, you have a creature that remains hidden while making that sound. This better hides the potentially troubling fact that the source of the advice is a bug. Third, crickets are omnipresent—they are found in many places where there are people. Lastly, the sound that creature produces can be thought to be in harmony with our psyche. When referring to the conscience, we often use the phrase "that little voice in my head." It is not a stretch to imagine that the voice could come from a tiny hidden creature that sees you, but remains unseen by you.

When my son Jeff was younger, I went through a period where I caught up with some of the longer children's literature I had bypassed as a child. I read him the entire original Oz series by Baum, Milne's *Winnie the Pooh* books, White's *Stuart Little,* everything by Roald Dahl, and Selden's *The Cricket in Times Square.* I'd come across Selden's book, written in 1960, as a library page working in the children's book section while going to college. What originally caught my attention were Garth Williams's illustrations, and I ended up reading the entire book in one day.

Unlike the character of Jiminy, this cricket, named Chester, was more of the "fish out of water" type. He had found himself in New York City after hopping into a picnic basket that made its way on a train from Connecticut. His singing brings joy to the people of Times Square and prosperity to the news-stand vendors who own him. Much of the tale is homage to the Asian custom of keeping crickets for their song. In fact, young Mario Bellini learns of this ancient practice from an old Chinese man, Sai Fong. Fong, a novelty shopkeeper in Chinatown, sets Chester up in an ornate red and green cricket cage with a golden spire.

George Selden got the idea for this Newbery Award–winning story when he heard a cricket chirping in a Times Square subway station. My guess is the cricket was a House Cricket, *Acheta domesticus.* It's the same species used in the cricket farm I wrote about earlier. This is an introduced insect that is more likely to be found indoors, and in more urban settings. It has long been known as the "Cricket on the Hearth." Chester, however, had to be a Fall Field Cricket, *Gryllus pennsylvanicus.* It's the species drawn by the illustrator, but also the only other large black field cricket likely to hop a train from Connecticut would be the Spring Field Cricket, *Gryllus veletis.* I ruled this one out because the adults die by early to mid summer. The call of *Gryllus pennsylvanicus* is a rich, repeated chirp. It's the sound most associate with cricket song. In the beginning of the book, when Mario first hears Chester, he describes the sound as "like a quick stroke across the strings of a violin." The songs of crickets and katydids have often been compared to the sound of a violin. This is not surprising, because like a violin, the insect's sound is produced by the drawing of one sound-producing device over the other. The scraper of the wing is the bow. The file is the strings. The rest of the wings amplify the sound and act as the body of the violin.

A good violinist will take many years to learn his or her craft. A cricket can break into flawless song the very moment it frees its wings from the skin of its last molt.

That song means different things to different people. In *The Reader's Handbook of Allusions, References, Plots* (1880) section under "Cricket," the Reverend E. Cobham Brewer writes, "Crickets bring good luck to a house. To kill crickets is unlucky. If crickets forsake a house, a death in the family will soon follow." However, a different publication from 1884 on the superstitions of Worchester says it is *unlucky* to have crickets in the house. I suppose both theories could be proven correct. Unlucky things can happen in a house playing host to crickets, and family members of people with no crickets in their home will eventually pass on. It is human nature to attempt to draw connections. We are always looking for the cause of the effect.

Eraldo M. Costa Neto set out to discover what some of those different meanings were in the village of Pedra Banca, Brazil.[1] To many of the villagers, the calling of a field cricket meant rain was on the way. A cricket in the room, however, would presage financial gain. In other cultures, a cricket in the room brings an omen of sickness or death. As for the katydids, a green-eyed katydid was good luck and a black-eyed katydid was an ill sign. Said one villager, "I am frightened when a katydid rests on me,

but I do not remove it from me. Otherwise my luck goes away."
Growing up, the kids in my neighborhood held that the praying
mantis was a harbinger of luck. In fact, we were of the belief
that there was a fifty dollar fine for killing one. I suspect the
praying mantises were behind that rumor.

As for the Cricket on the Hearth I made reference to earlier,
they have been associating with the likes of us humans since we
began dwelling in shelters. House, or Hearth, Crickets don't
do very well in the winter and find our warm homes suit them
nicely. As unpaying and uninvited tenants, they have been seen
as both moving instrumentalists and pests. They are presag-
ers of good luck and foretellers of bad. Charles Lester Marlatt
summed this up in a USDA bulletin in 1896:

> No insect inhabitants of dwellings are better known
> than the domestic or hearth crickets, not so much
> from observation of the insects themselves, as from
> familiarity with their vibrant, shrilling song notes
> which, while thoroughly inharmonious in them-
> selves, are, partly from the difficulty in locating the
> songster, often given a superstitious significance and
> taken, according to the mood of the listener, to be
> either a harbinger of good and indicative of cheerful-
> ness and plenty, or to give rise to melancholy and to
> betoken misfortune.[2]

In other words, the benefit gleaned from a singing House
Cricket is dependent upon the listener's state of mind. It also
depends upon whether or not the crickets have been chewing
away at your clothes or upholstery, which they have a tendency
to do. One must be careful of retaliation, though, because of
the belief that the relatives of a slain cricket will exact their
revenge by cutting up the garments of the offender.

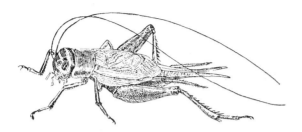

Artist: L. Joutel, from "Bulletin American Museum of Natural History [Vol. 6],
Article XII—Descriptive Catalogue of the Orthoptera Found within Fifty Miles

One of the earliest reports of these non-native House Crickets in North America came from Peter Kalm in 1749. Kalm, a Swedish naturalist who set out to explore the New World, wrote that even then they were abundant in Canada, "especially in the country where these disagreeable guests lodge in the chimneys nor are they uncommon in the towns. They stay here both summer and winter and frequently cut clothes in pieces for pastime."[3]

I do love the suggestion that a cricket would do *anything* "for pastime."

In 1845 Charles Dickens published a novella called *The Cricket on the Hearth*. The book begins with showdown between a whistling kettle and a chirping cricket.

> And here, if you like, the Cricket DID chime in! with a Chirrup, Chirrup, Chirrup of such magnitude, by way of chorus; with a voice so astoundingly disproportionate to its size as compared with the Kettle (size! you couldn't see it!) that if it had then and there burst itself like an overcharged gun, if it had fallen a victim on the spot and chirruped its little body into fifty pieces, it would have seemed a natural and inevitable consequence, for which it had expressly laboured.

The kettle never stood a chance. It was up against a magical being, a "potent spirit" who could take the form of a faerie and whose song presaged good tidings. Dickens's cricket was similar to Colladi's Talking Cricket, imbued with the ability, and inclination, to look over the stumbling humans. At times crickets were even assigned to keep the egos of other animals in check, as seen in the old tale shared by many western Native Americans. It is the cricket that shears the bushy tale of the vain opossum, thus sentencing the mammal, and its descendants, to life with an ugly, naked tail.

Today's House Crickets are held in lesser esteem. They are used as bait by fishermen and are the hapless species most often seen within the cages of reptiles and amphibians. They're not in there to offer advice or bring luck to their herpetological cage mates, either. It is a House Cricket's unfortunate lot in life to be raised solely as pet food or fish bait.

There is a bit of cricket lore that is rooted in fact; it involves the Thermometer Cricket's ability to tell the outdoor temperature. The species is *Oecanthus fultoni,* also known as the Snowy Tree Cricket—"snowy" because of its very pale, ivory-green color. They are an arboreal species found throughout most of the country, though absent from the Southeast. If you haven't heard it outdoors, chances are you've heard it on television or in the movies. Because its rich, steady chirp lends itself to adding a country summer night ambiance to an outdoor scene, its call is often dubbed in from a studio recording. It is the Kookaburra of the crickets, the former, the bird often heard in the background of early movies taking place in Africa. However, unlike the Snowy Cricket, which tends to be more correctly placed where it belongs, the Kookaburra is an Australian bird never found in Africa.

Most crickets and katydids speed up the frequency of their call as temperature increases, and slow it down as it decreases. The

Snowy Tree Cricket, however, has long been known to maintain a steady pulse in consistent unison with minor temperature fluctuations. There are a number of formulas out there, but the one I've come across more frequently requires you to add forty to the number of chirps counted in a thirteen-second period. This should equal the outdoor temperature; give or take a degree or two. I've tried this and can personally testify that it works.

While crickets have continued to surface in our literature and folklore, we hear less about their cousins the katydids.

We do know that in ancient China, katydids were used as aphrodisiacs, perhaps in light of their ability to lay hundreds of eggs. In the United States, many see the Common True Katydid as the harbinger of the first frost. The first scratch of their wings signals that the first blast of icy cold is just twelve weeks away. When it begins to call in the late afternoon, we're down to six weeks. The katydids do call earlier in the evening as the season progresses, perhaps in desperation to get done what needs to be done before the season, and their life, has ended. This axiom has a prophetic, folkloric feel to it, but it's just simple math. If the Common True Katydid begins calling around midsummer, which it does, then one just has to count the weeks when the first frost usually arrives. In much of central and northeast United States, that ranges between mid-September and mid-October.

The word *katydid* is a mnemonic of the call of the Common True Katydid, *Pterophylla camellifolia*. Its name was assigned in 1775 by a student of Linnaeus—Johann Sebastian Fabricius. If you translate the scientific name, you end up with "Wings shaped like a camellia leaf." Camellia is the Latinized name of the Reverend Georg Kamel, a late-seventeenth-century botanist and missionary to the Philippines. The wings of the insect described by Linnaeus's student could very well be compared to the Reverend Kamel's leaf.

Sue Hubbell, in her book *Broadsides from the Other Orders* (1993), touches upon the origin of the word katydid.[4] She wonders if the "katy" in katydid had a dual origin. Hubbell noted that *The Oxford Dictionary* includes the word *wanton* under the definition for *katy*. This would certainly tie in with an old tale she recounts, which originated from the hills of North Carolina. It tells of a handsome young man who chose one sister over the other. He and his new wife are both found dead in the honeymoon bed, and the surviving sister, Katy, is the prime suspect. The insects in the trees obviously took an interest in this and can be heard debating as to whether or not Katy did it.

Naturally, if you have an insect that repeatedly makes a statement, that being "Katy did," there's bound to be the crafting of poem and tale to explain exactly what Katy, did in fact, do. In 1831, Oliver Wendell Holmes had fun with the name in his poem "To an Insect." In one stanza he asks:

> Oh, tell me where did Katy live,
> And what did Katy do?
> And was she very fair and young,
> And yet so wicked, too?
> Did Katy love a naughty man,
> Or kiss more cheeks than one?
> I warrant Katy did no more
> Than many a Kate has done.

Hubbell looked at the name from another direction, though, and discovered that the first published reference to this insect came from botanist John Bartram in 1751. Bartram had used the word *catedidist*. Hubbell observed that *cate-* had Greek origins for "resound, to din in one's ears." I was able to find that Bartram passage in a journal of his trip to New York. It was published in London and reached the colonies by 1752.

Meadow Katydids (Conocephalinae)
Slender Meadow Katydid
(*Conocephalus fasciatus*), female from Maryland meadow.

Meadow Katydids (Conocephalinae)
Short-winged Meadow Katydid
(*Conocephalus brevipennis*), female from Connecticut yard.

Meadow Katydids (Conocephalinae)
Saltmarsh Meadow Katydid
(Conocephalus spartinae), male from Connecticut salt marsh.

Meadow Katydids (Conocephalinae)
Red-headed Meadow Katydid
(Orchelimum pulchellum), female from weeds outside Savannah, Georgia, motel.

Meadow Katydids (Conocephalinae)
Common Meadow Katydid
(*Orchelimum vulgare*), female from a fen in Connecticut.

Meadow Katydids (Conocephalinae)
Black-legged Meadow Katydid
(*Orchelimum nigripes*), female from a Connecticut meadow.

Meadow Katydids (Conocephalinae)
Gladiator Meadow Katydid
(*Orchelimum gladiator*), male from a fen in Connecticut.

Meadow Katydids (Conocephalinae)
Seaside Meadow Katydid
(*Orchelimum fidicinium*), from a salt marsh in Connecticut.

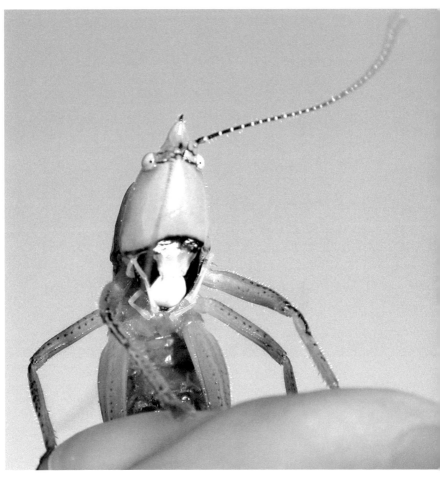

Coneheads (Copiphorinae)
Davis's Conehead
(*Belocephalus davisi*), male from a swamp in Georgia.

Coneheads (Copiphorinae)
Round-tipped Conehead
(Neoconocephalus retusus), female nymph from Virginia.

Coneheads (Copiphorinae)
Round-tipped Conehead
(Neoconcephalus retusus), adult female from roadside weeds in Connecticut.

Coneheads (Copiphorinae)
Eastern Swordbearer
(Neoconocephalus ensiger), male from a Connecticut meadow.

Coneheads (Copiphorinae)
Long-beaked Conehead, aka Slightly Musical Conehead
(Neoconocephalus exiliscanorus), from a Connecticut fen.

Coneheads (Copiphorinae)
Nebraska Conehead
(Neoconocephalus nebrascensis), Virginia female—
head to ground, mimicking grass blade.

Coneheads (Copiphorinae)
Nebraska Conehead
(Neoconocephalus nebrascensis), male from edge of a cornfield in West Virginia.

Coneheads (Copiphorinae)
Marsh Conehead
(Neoconocephalus palustris), female from edge of a pond in Maryland.

False Katydids (Phaneropterinae)
Rattler Round-winged Katydid
(*Amblycorypha rotundifolia*), male in a Connecticut vegetable garden.

False Katydids (Phaneropterinae)
Oblong-winged Katydid
(Amblycorypha oblongifolia), female in weeds along a Virginia road.

False Katydids (Phaneropterinae)
Greater Angle-wing
(*Microcentrum rhombifolium*), female from a Connecticut yard.

False Katydids (Phaneropterinae)
Northern Bush Katydid
(*Scudderia septentrionalis*), male from a Connecticut yard.

False Katydids (Phaneropterinae)
Fork-tailed Bush Katydid
(Scudderia furcata), male nymph from a Connecticut yard.

False Katydids (Phaneropterinae)
Fork-tailed Bush Katydid
(Scudderia furcata), female from a Connecticut yard.

False Katydids (Phaneropterinae)
Curve-tailed Bush Katydid
(*Scudderia curvicauda*), male from a Connecticut meadow.

False Katydids (Phaneropterinae)
Treetop Bush Katydid
(*Scudderia fasciata*), female from a Connecticut bog.

True Katydids (Pseudophyllinae)
Common True Katydid
(*Pterophylla camellifolia*), male from Connecticut.

Shield-backed Katydids (Tettigoniinae)
Protean Shieldback
(*Atlanticus testaceus*), male from leaf litter along a Connecticut trail.

Shield-backed Katydids (Tettigoniinae)
Roesel's katydid
(Metrioptera roeselii), male from a Connecticut meadow.

Mole Crickets (Gryllotalpidae)
Northern Mole Cricket
(*Neocurtilla hexadactyla*), female from the bank of a Connecticut pond.

Quiet-calling Katydids (Meconematinae)
Drumming Katydid
(*Meconema thalissinum*), female from a Connecticut yard.

Field Crickets (Gryllinae)
Spring Field Cricket
(*Gryllus veletis*), male nymph from a Connecticut yard.

Field Crickets (Gryllinae)
Fall Field Cricket
(*Gryllus pennsylvanicus*), male from a Connecticut yard.

Field Crickets (Gryllinae)
Spring Field Cricket
(*Gryllus veletis*), female from a Connecticut yard.

Field Crickets (Gryllinae)
Sand Field Cricket
(*Gryllus firmus*), female found along a Connecticut river.

Field Crickets (Gryllinae)
Japanese Burrowing Cricket
(*Velarifictorus micado*), male found under car in Maryland.

Ground Crickets (Gryllinae)
Tinkling Ground Cricket
(*Allonemobius tinnulus*), male from leaf litter in yard.

Ground Crickets (Gryllinae)
Striped Ground Cricket
(*Allonemobius fasciatus*), female from a Connecticut fen.

Ground Crickets (Gryllinae)
Southern Ground Cricket
(*Allonemobius socius*), female from side of road in South Carolina.

Ground Crickets (Gryllinae)
Sphagnum Ground Cricket
(*Neonemobius palustris*), male from a quaking bog in Connecticut.

Ground Crickets (Gryllinae)
Carolina Ground Cricket
(Eunemobius carolinus), male from a Connecticut lawn.

Bush Crickets (Eneopterinae)
Jumping Bush Cricket
(Orocharis saltator), male from Connecticut.

Bush Crickets (Eneopterinae)
Restless Bush Cricket
(Hapithus agitator), male from along a Connecticut stream.

Bush Crickets (Eneopterinae)
Short-winged Bush Cricket
(Hapithus brevipennis), female from a swamp in Georgia.

Bush Crickets (Eneopterinae)
Restless Bush Cricket
(Hapithus agitator), female from tall weeds in New Jersey.

Scaly Crickets (Mogoplistinae)
Forest Scaly Cricket
(Cycloptilum trigonipalpum), male from phragmites in New Jersey.

Tree Crickets (Oecanthinae)
Two-spotted Tree Cricket
(Neoxabea bipunctata), female in a Connecticut yard.

Tree Crickets (Oecanthinae)
Two-spotted Tree Cricket
(Neoxabea bipunctata), male sipping moisture in a
Connecticut yard.

Tree Crickets (Oecanthinae)
Four-spotted Tree Cricket
(Oecanthus quadripunctatus), male from Delaware showing antennae base markings.

Tree Crickets (Oecanthinae)
Black-horned Tree Cricket
(Oecanthus nigricornis), male from a Connecticut meadow.

Tree Crickets (Oecanthinae)
Pine Tree Cricket
(Oecanthus pini), female in a white pine in Connecticut.

Tree Crickets (Oecanthinae)
Fast-calling Tree Cricket
(Oecanthus celerinictus), male from side of road in Georgia.

Tree Crickets (Oecanthinae)
Broad-winged Tree Cricket
(*Oecanthus latipennis*), female in a meadow in Ohio.

Trigs (Trigigoniinae)
Say's Trig
(*Anaxipha exigua*), male nymph from North Carolina.

Trigs (Trigigoniinae)
Say's Trig
(Anaxipha exigua), female (long-winged form) in a meadow in Connecticut.

After having supped on venison shot by the Indians earlier, he wrote, "It was fair and pleasant and the great green grasshopper began to sing *(Catedidist)* these were the first I observed this year."

Just thirty-three years later, John Smyth had referred to this same species as "katy did." I located that reference in Smyth's book *A Tour of the United States of America* (1784). He wrote, "There is a very singular insect on this island [Long Island, NY], which I do not seem to have observed in any other part of America. They are named by the inhabitants here katy dids." It is entirely possible that Bartram's "catedidist," as pronounced by those Long Island locals, was heard as "katy did." Words have a way of evolving in their travels from mouths to ears. It is not until they are repeatedly shown in a consistent written form that they become *somewhat* rooted.

It is interesting to note that in places beyond the range of *Pterophylla camellifora,* cultures had to come up with different names for their own "katydids." They did not have the "kaytee-did" call to work with. In the United Kingdom, they are known as bush crickets. In Spain and Portugal, katydids are esparanzas. *Esparanza* means "hope." The color green, worn by most esparanzas of the area, is a symbol of hope, hence the attribution. In Germany they are laubheuschrecke, which basically means leaf locust. Heuschrecke is an interesting word for locust, or grasshopper; it translates as "hay frightener."

I was in Cozumel, Mexico, in 2008, where I came across a man who appeared to be making a living on katydids; not real katydids, but ones made of palm fronds. He could weave one together in just under a minute, and would hand them out to tourists on the beach. Naturally he counted on (but did not insist upon) a donation. These were perfect replicas of bush katydids, with long hind legs and leaf-shaped wings. I tried to tell him what we called them in the States, but he was unfamiliar with the term, preferring instead, *saltamontes,* which simply means grasshopper. I think most people around the world, unfamiliar with the distinction between the two families, settle on the word *grasshopper* for a katydid they may have come across.

There is also a variety of colloquialisms regarding this bug. *The Dictionary of American Regional English* lists *cha-cha, night-did, chidderdiddle, frost bug, kittledee,* and *sniddydid.* One of my favorites of those I've come across is used in North Carolina—*chatteracts.* They've also been called katyDIDN'Ts:

Woven palm-frond katydid.

> We landed amidst a file of soldiers . . . and a jabber-
> ing of voices which can only be rivaled among the
> Katydids and the Katydidn'ts of Connecticut.[5]

It should be noted that the Common True Katydid does
sometimes add another syllable to the call, making "katydidn't"
a viable translation.

Crickets seemed to exude a certain magical quality in our folklore. Katydids, less so. This may have something to do with their
call, which carries less of a musical pitch than that of the crickets.
Pleasant as those staccato sounds are to my own ear, their notes
tend to be drier, with a scratchier quality. Katydids also forgo
our homes as shelters, preferring to be among the leaves they
resemble. This puts them less frequently in our physical path.

That is not to say that katydids were completely spared from
fanciful notions. An old Cherokee legend tells of two hunters
camping in the woods. One of the hunters mocks a singing
katydid, saying that it doesn't even realize it will be dead by the
end of the season. The katydid accepts that bit of fact, but tells
the hunter that he, in fact, will be dead before the next night.
True to the prediction, the hunter is killed by an enemy. The
lesson? Don't sass the katydids.

I found another somewhat magical account of katydids in
The Life of North American Insects, published in 1859 by Benedict
Jaeger. He shares, with unhidden incredulity, evidence that in
some areas, science had yet to pierce through myth. In fact, he
went out of his way not to mention that writer by name (so as
not to embarrass him, I presume):

> A certain traveler, in a work on America published
> several years ago, related the most absurd stories in
> regard to these insects. He said that on this Continent
> an animated insect often changes itself into a lifeless

plant, by putting its feet into the ground and allowing them to take root, when they actually become the stems of a foliated plant. That leaves are sometimes changed into insects with a distinct head, throat, abdomen and legs. No one, he says, can doubt these facts, as there are in Brazil thousands of witnesses who are ready to prove that they have often observed these phenomena.

The mid-nineteenth-century belief that a katydid can become a plant, or a plant a katydid, is easy to forgive, especially in light of the fact that Darwin's theory of natural selection had come out the very same year as Jaeger's book. It would take some time for this theory, which gave a better reason for resemblance between leaf and bug, to reach out into the world.

Nearly a hundred years prior, Linnaeus described a katydid he may have very well seen in his gardens while growing up in Sweden. In fact, it is possible that either he, or someone he knew, had accepted its unwittingly offered medical services. In 1758 he described the species *Decticus verrucivorus. Decticus* is derived from the Greek adverb meaning "biting." *Verrucivorus* comes from the Latin root word *verruca,* which means "wart." When you add "vorus" to a name, you are stating that the animal you are describing devours what is named by the preceding word.

Wart Biters, as they are commonly known, were once used as nature's wart removers. They are armed with large and powerful mandibles, which were put to use to chew off the unwelcome growths. The katydid would be held in position so that its mastication would be confined to that specific area. A small wart could be removed with just a few chomps. Larger warts would present more of a challenge, not for the katydid, who would happily munch away for as long as there was flesh to chew, but for the person on the other end of those mandibles!

This species occurs throughout Western Europe, though in England they have become so rare that they are one of the 391 species protected by the United Kingdom Biodiversity Action Plan.

There is a katydid in the western United States that falls on the other end of the spectrum. Mormon Crickets (a kind of shield-back katydid) would be considered anything *but* rare! When conditions are right, they gather in large, crawling swarms numbering into the millions.

The name "Mormon Cricket" comes courtesy of some of the earliest stories of this insect, as they laid waste to the fields of the Mormon settlers of Utah. There are several wonderfully colorful reports of this march published in 1889 by Hubert Howe Bancroft. In his *History of Utah,* he shares the ravaging effects, both psychological and material, these insects had on those early settlers. In one such passage he writes about an invasion of Mormon Crickets in 1848:

> They came in a solid phalanx, from the direction of Arsenal Hill, darkening the earth in their passage. Men, women, and children turned out en masse to combat this pest, driving them into ditches or on to piles of reeds, which they would set on fire, striving in every way, until strength was exhausted, to beat back the devouring host. But in vain they toiled; in vain they prayed the work of destruction ceased not, and the havoc threatened to be as complete as was that which overtook the land of Egypt in the last days of Israel's bondage.

Just when all seemed lost, there appeared "myriads of snow white gulls." They descended upon the dark horde, gorging themselves on the easy feast. They left behind great piles of chitinous pellets, those remnants regurgitated in the manner of most insectivorous birds. The saviors have since been determined to be California Gulls. In 1955 the California Gull was named the state bird of Utah, in recognition of the fact that it had been a revered bird since that 1848 event.

One year prior, Lorenzo Young (brother of the famous Brigham Young) recorded this migration, but took special interest in how the Indians took advantage of it. They made a large corral with sago brush and greasewood, and drove the crickets into the enclosure. The brush fence was then set on fire, cooking the insects. Afterward they rubbed off the wings and legs and separated the "meat." Young estimated there was an ounce or two of fat in each cricket.

Mormon Crickets feed on herbaceous plants, vegetables, and other cultivated crops, making them a terrible nuisance to farmers. They also eat each other, which is not an uncommon trait for shield-backs. A 2006 study revealed that the marching insects were deprived of two essential life elements: protein and salt.[6] While these katydids may be low in these elements, their bodies are not completely void of them. This makes each neighbor a tempting solution to their problem. A picture comes to mind of one of those clichéd cartoon devices where two starving men see each other as a hotdog or hamburger. The authors write, "As a result, the availability of protein and salt in the habitat will influence the extent to which bands march, both through the direct effect of nutrient state on locomotion *and indirectly through the threat of cannibalism by resource-deprived crickets approaching from the rear.* The crickets are, in effect, on a forced march."

This gives a whole to meaning to the term "move it or lose it!"

Grasshoppers, which share the order Orthoptera with the crickets and katydids, usually catch the brunt of our ire when it comes to our crossing paths. This is always in relation to

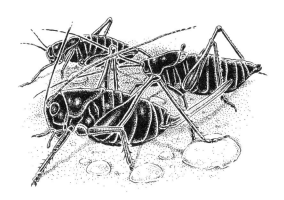

Mormon Crickets (*Anabrus simplex*) on the march.

the crop damage for which they are responsible—not all of them, but a select handful of species. Katydids have generally kept beneath the radar, although they can be the bane of fruit orchard owners. Katydids love fruit! Aside from the Mormon Crickets, though, they don't build up in swarms, as do some of the locusts. Their lower population densities allow the farmers to deal with them as individuals, as opposed to battling a giant entity composed of thousands or millions of insects.

The crickets and katydids tend to mind their own business. They keep out of our way, hiding in the leaves or the dark nooks and crannies of the outdoors. Most cultures welcome their nocturnal song, despite having never seen most of the singers. We hear them when we're not listening. They call in our subconscious, and sometimes, even, *as* our conscience.

eight

A Blade within a Sea of Grass

Adventures in Hunting Katydids and Crickets

I'm beginning to feel the early twinges of frustration. How, one may ask, can anyone feel even a modicum of displeasure while standing in the middle of a thirty-five-acre wildflower-filled fen? It doesn't get more bucolic than that. I tell myself to relax. Besides, I'm just hunting a bug, and to paraphrase an old fishing cliché, "A bad day's hunting bugs is better than a good day's work."

However, over the last decade my job has more or less revolved around hunting bugs, so a bad day of hunting bugs *is* a bad day's work. I hear the insect ticking away in the sedges, *but I just can find it!*

Of all the insects, the ones that truly try my patience are the larger meadow katydids, and that's saying something. You need little patience to find most of the singing insects. Insects that produce sound must find that fine balance between allowing the insects within their species to find them, and remaining hidden from the hordes of creatures that want to eat them. Although there are human cultures that do add a healthy serving of insects to their diet, I'm not among them. I have no desire to eat this, or any other, katydid. They don't know that, though. Therefore, as soon as it detects my presence, a singing katydid will go silent until the perceived threat is gone.

Perhaps, to be fair, I should not consider myself merely a perceived threat. If I find the insect, I do intend to catch it and take it home. It would then reside for a week or so in my studio so I can record its call and get to know it a little better. So no, I won't eat it, but I'm sure that, if given the choice, the insect would opt to stay where it is.

The particular katydid I'm stalking is *Orchelimum vulgare,* or the Common Meadow Katydid. It's a rugged-looking little creature, about the size of my pinky. They are the color of grass—blue-green and reddish brown, with tapered wings that help them blend in with the pointy monocots that surround them. The stridulary areas of the outer wings have the characteristic double humps of this family that when rubbed together create a series of ticks or a sustained whir. These insects inhabit meadows throughout the North and South, from the central United States to the Atlantic Coast, where they feed on seeds and, occasionally, other insects.

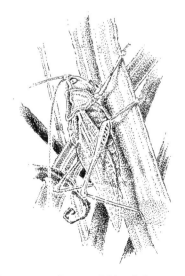

Common Meadow Katydid *(Orchelimum vulgare).*

Common Meadow Katydids fall in the genus category of Orchelimum, or larger meadow katydids. This is to distinguish them from the Conocephalus, or smaller meadow katydids, which are represented in this fen by two species. One of those species, the Short-winged Meadow Katydid, is also calling. The sound is similar to the tick and whir produced by the Common Meadow Katydid, but softer. They are very easy to find and catch, because they frequently opt for hopping out of the way instead of the Common Meadow Katydid's method of freezing and hiding.

I'm not the only thing this Common Meadow Katydid has to fear, of course. Birds, wasps, mantids, rodents, amphibians, and reptiles all are the stuff of nightmares for unwary crickets and katydids. Aside from a few of the more tenacious species of katydids that can give a good strong nip, when it comes to an active defense against predators, there's not much this group of insects have going for them. What they've become adept at is remaining unseen and choosing when to be unheard.

Part of staying beneath predators' radar involves the ability to hear them coming. Katydids hear very well. Their ears are located on the upper part of the two front tibiae, right where they meet the femora. Simply put, they're just below the front elbows. The ear is visible to the naked eye as a small, oval indentation, often contained within a flattened surface. It is called the *tympanum,* or *tympanic membrane,* and is a relatively simple structure compared to our own hearing apparatus. Theirs is similar to the tympanum visible just behind the eyes of frogs. A thin drumlike membrane stretches across the opening to catch the air vibrations that make up sound.

Having one's ears on one's elbows may seem like an inconvenience. Specific needs and circumstances call for different shapes, though, and a katydid that lives in a habitat surrounded by sound-dampening structures, such as leaves, stems, flowers, and branches, gains an advantage in being able to hold two ears as far from the body as possible. In addition to creating a larger field from which to collect sound, the ability to move the ears independently allows them to pinpoint the source of a sound. This facility is akin to the advantage in depth perception gained by the use of two eyes instead of one. One eye will show you the object in front of you. The second eye helps determine how far it is from you. When it comes to hearing, each ear will collect the sounds made within their range. However, unless that sound is equidistant from the two ears, one ear will perceive the sound at a louder volume than the other. If the listener can adjust the positioning of those sound-trapping organs, the ability to isolate the source of that sound is increased.

Therein lies my frustration.

The Common Meadow Katydids aren't giving me a fighting chance this afternoon. Each time I lock on one in the distance, it hits the mute button as soon as I get to within about twenty feet of it. As if this wasn't enough, the Orchelimum have another weapon in their arsenal of cryptic strategies. Upon seeing a threat, they shimmy over to the opposite side of the stem, putting that stem between them and the threat. The body is then stretched out, long and parallel to the stem, to reduce their profile. The legs and antennae are also held in this vertical position. They will stay like this, in silence, until the danger has passed. This works very effectively for most of the species within this genus. Part of the reason it works so well is that their coloration blends in so nicely with the surrounding environs. There is one meadow katydid, however—the Red-headed Meadow Katydid—that provides less of a challenge. I first came across these beautiful creatures just a few dozen feet from a hotel room door in Savannah, Georgia.

I had opened the door to ask a couple of guys outside my

room to find another place to hang out and drink beer. They were the only other people in the entire place, and the manager chose to put them next door to me. I suppose it was to make cleaning the rooms easier in the morning. They were getting a bit loud, and repetitive. The latter was most annoying because I didn't need to hear for the fifteenth time about the canine lineage of someone named Larlene (or something to that effect, it came out more garbled each time). They apologized and offered me a can of beer, which I might have taken them up on, had I not been busy writing. I declined, and they went back to where they left off. Just as I went to close the door and go back into my room, I heard a buzz in the weedy lot just across from me. It was a steady, sizzling sound that seemed to trip over a "tick" every second or so. I walked toward that sound and saw the striking redheaded singer. The sun was just beginning to set, and the cherry-red noggin literally glowed in the low, diffused sunlight. It then performed its "cryptic shimmy" around to the far side of the stem. Other calls emanated from that same weedy area, and as I looked around I saw the beaming heads of two more males, and three females. There has to be a reason why this particular genus evolved the striking beacons that gave them away to this human stalker, although I am confident none of their evolved attributes or disadvantages were in response to my species. I looked back toward my room, where I had left the door open in my haste to get to the field. The drunks were staring at me, and I knew they were going to ask me what I was doing when I walked back. I also knew any honest answer I gave them would just leave them more confused, so I told them I was writing a book about grasshoppers. "Bait!" said one of them, which was repeated by two of the others. This really seemed to amuse them. I smiled, made the universal sign of casting a lure, and went back inside to get my camera. Hey, they weren't wrong. In fact, I'd bet those red-headed 'soppers would have made great fishing bait. I can picture them spinning around on the line, trying to hide from a fish.

I went back to the field and easily located a couple of males and females. After taking some pictures, I went back to my room. On my way, one of the guys made an invisible cast with his invisible rod and reel. "Bait! Good bait!" he said. "Can't beat it," I said.

For some, the thought of a grown man watching bugs, for the sake of watching bugs, is an unfathomable concept. Maybe I misjudged this group and missed an opportunity to share something interesting. I don't know. In this case instinct suggested otherwise.

That hiding technique employed by the larger meadow katydids could be what earned their genus the Greek name of Orchelimum, meaning "Meadow Dancer." You go to the right. The bug goes left. You go left. The bug goes right. The two of you can go around and around in complete circles as it continues to take evasive action behind the grass stem. It is a dance, of sorts, perhaps most closely resembling a waltz.

Skills learned in stalking meadow katydids can be applied to hunting most of the other singing insects. It's like learning to track flying birds with your binoculars by practicing on darting hummingbirds. Once you nail that skill down, you can pretty much track anything. The hunt for the Orchelimum can be considered your basic training.

It begins by learning how to pinpoint their general location. The first step is the most simple: stand and listen. Relax your ears much in the way you would relax your eyes to take in a wider swath of your environment. In reality, you're really just widening your awareness. Listen for every sound you can hear in the meadow. Widen your hearing range by slowing turning your head back and forth. As you do so, some sounds will fade and new ones will present themselves. You may hear tree crickets trilling in the flowers and among the leaves of the low-growing plants. Some may be in the surrounding trees. Ground and field crickets will likely be chirping or trilling from the meadow floor, among the tangle of roots and beneath the dead leaves. There's rarely a long stretch of time before an airplane flies overhead. The more one opens one's ears, the more one realizes how few areas are free from the sound of airplanes. Or barking dogs.

If the meadow katydids are your quarry, listen for their isolated ticks, or series of them. The calls will be coming from the grasses and sedges. If you hear one, there will likely be several more within your audible range.

Once you're comfortable with your initial aural survey of the area, you will need to fine-tune your hearing. This is accomplished by cupping the back of your ears with your hands. It's an age-old sound amplification technique used by young and old throughout the world. It's something we've learned to do as children by watching others do the same, and it's an action that for many of us has become a natural reflex. Slowly tilt your head up and down and attempt to distinguish between the calls coming from up in the surrounding trees, and the calls lower in the meadow. This is especially fun to do at night, when more of the insects are calling from their arboreal perches. Without the aid of your hands, the mixed calls of the night can be overwhelming. As soon as you cup your ears to focus in on the different calling zones, you gain the ability to separate who is calling from where. This technique teaches your brain to focus on the individual and specific sounds.

In an open meadow, you are really mostly concerned with scanning side-to-side. As your head swivels along the horizontal plane, certain sounds become louder as your hearing zone passes over the caller. Fine-tune your positioning by adjusting your hands, and the direction your ears are facing, until the call is at its loudest. I learned a great term for this from a cuckoo-clock repairman. He showed me that in order for these kinds of clocks to function at their best, they need to be tipped in the ideal position on the wall. To find that position, you put an ear against the clock to listen to the ticking while moving it ever so slightly from side to side.

What you're listening for is the strongest and steadiest sound of ticking, which he called the "sweet spot." So when scanning the meadow, hands behind ears, you are fine-tuning your focus until you find that sweet spot. Once you have the general direction, you can gauge the distance by tipping your head up and down. The farther the source of the call, the more your head will be tilted up when you settle on the sweet spot. Then it's time to move in.

I usually start with either the closest insect or the one with the fewest obstacles between us. Dead leaves on the ground and branches reaching across the route are the stalker's nemesis. It's hard to judge the distance at which a particular insect will notice you, or act on your presence. I usually get a sense of it after spooking a few. When you feel that it's time to begin to mask your sound and movements, you will shift into stealth mode. Begin your approach toward the source of the call by slowly sliding one foot out in front of you. Keep the rest of your limbs and body as motionless as possible. Then slowly pull your rear foot forward, which will bring the rest of your body with it. Keep your movements fluid. If there is a breeze rocking the plants, you may want to rock gently with them. I suppose this could be part of that meadow dance I wrote about earlier. I actually learned this from watching the round-headed katy-dids. When they sense danger but still feel the need to move, they rock like a leaf in the wind. The whole body moves back and forth as each leg is tentatively extended and set down. Some birds employ a similar blending in action. Waterthrushes, which spend a lot of time at the edges of water, will bob their hindquarters up and down, mimicking the movement of ripples and small waves at the shore. Moving in unison with the flora can have two very different results: One, it will make your shape just one more piece of the surrounding habitat, which may be anything *but* still. This could add some forgiveness to any other minor random movement emanating from your body. Or two, it will immediately give you away to the creature you're stalking. I suppose there are times when people make unconvincing plants. Whether you become one with the wind, or not, the goal with the first step is to move only the lower part of your body, which should mostly be concealed by the understory of grass and other low plants. The rest of your body catches up as the second foot follows and is set quietly on the ground.

After every step or two, stop and listen. If the insect is still calling, chances are you haven't been seen or heard yet. If it has stopped, don't resume moving until it calls again. While you're waiting for your original caller to begin singing, listen for others nearby, should you need to abandon that one. Giving up on a particular individual you may have been stalking for fifteen minutes is a tough call to make. You've already invested the time in this one. However, there's a chance that if it remains silent for too long, it could be because you are right on top of it. If this is the case, it will stay hidden until you move on. Of course, as soon as you give up and move toward the next one, it will start calling again. I am not above swearing at a bug.

I often find it necessary to hold my hands behind my ears to stay on target. Your elbows should be held close to your sides to maintain a minimum profile. Again, it is important to remember to keep all movements slow and fluid, especially when your hands are breaking out from your general countenance. That's a rule for any kind of stalking. Waving hands and arms attract attention. Just think about how we humans use this movement to communicate with one another. If we are desperate to signal danger, or just wish to get someone to look in our direction, we wave our hands like signal flags.

The Common Meadow Katydid I find myself stalking on this day has obviously seen or heard me. After another five minutes of standing still and waiting, I'm ready to go to the next step. Figuring that the moment of surprise is lost, I get down on my knees to scan the grasses. I am now within its world, seeing the surroundings from its vantage point. What I look for are the antennae. This katydid's body and wings give it the look of a blade of grass, so chances are that unless it's moving, my eyes will pass right over it. The antennae, however, are more difficult to conceal and will sometimes be sticking out from behind a stem, and may actually be moving. If it were nighttime, I would be running my flashlight up and down the stems. Moving the beam of light in this direction can sometimes help find breaks in the vertical patterns of the grasses. Because it is daytime, I simply move my head up and down the stems, hoping to find that same break in the pattern. Having found none, I am left with three options: (1) Wait some more. (2) Move on to the next calling insect. (3) Sweep the area with my net.

This last option must be last, because if you miss, you blow any chance of finding that insect. Net sweeping, however, is a viable technique for catching all kinds of insects. It involves swinging your net through the flora, with the hope of scooping up the unseen insects hiding within. It's done with quick side-to-side motions to catch them unaware. If you move too slowly, they will scuttle down and hunker in the bases of the plants.

I've decided that for this particular insect, a net sweep is in order. I swing the net back and forth in the vicinity in which it was last heard. This inevitably scoops up all kinds of vegetation, partially filling the bag. I tap on the bottom to loosen up the debris. Various insects climb and fly out of the opening. One is a Common Meadow Katydid, but a female, as told by the presence of an ovipositor. It's not unusual to catch females of any katydid in this manner. They often gather around the calling males. I put her in a holding jar, with the hope that she will be joining a male in its terrarium, should I be lucky enough to actually capture one. A few more scoops prove unfruitful in that pursuit. It's time to move on to the next one I hear calling. It sounds as if it's about twenty feet away.

Once I've cut that distance in half, I look for the calling insect. From this distance I stand a fairly good chance of spotting it, especially with the aid of the binoculars, which I usually bring with me. I don't think most entomologists are in the habit of wearing binoculars, but a good close-focus pair is an excellent tool for observing insects. A calling Common Meadow Katydid is a moving target. While the body remains still, its outer wings vibrate and the antennae often whip around. If I can spot it from this distance, I may be able to catch it. I cup my hands behind my ears to zero in on the call, and then raise my binoculars to scan the area where it seemed loudest. There it is—ticking away. The wings move in a blur. I memorize its location. It helps to look for other plants in relation to the one upon which a particular insect is perched. I know full well that when I approach this katydid, it will spin around and hide. All I will have to go by is my memory of the exact stalk it was calling from, which looks like every other stalk in this meadow.

As predicted, the katydid scuttles around to the other side of the stalk as soon as it sees me. "Gotcha," I say, and make my way to where it's hiding. I can't see it, but I know exactly where it is. Part of the fun in finding these insects is in taking their picture where they are found. The unposed photograph puts the insect in context with where it lives. These creatures are very much a part of the habitat. They evolved to look like their surroundings; to feed on their surroundings; to mate and lay eggs in their surroundings; and to hide in their surroundings. That latter adaptation would make it nearly impossible to take a picture of it. However, I know a little trick. With the camera in my right hand focused on the stem and ready to click, I reach around in a wide arc with my left, coming up behind its hiding place—the stem. The katydid shimmies around the stalk to hide from my hand, bringing it into full view. I take a couple of photos and then scoop it up in my jar. When I get home it will reside in a nice, comfy, grass-filled terrarium. It will even have a female to sing to. Then in a week or so, after I've recorded its call and taken a few more photos, they will both be returned to their fen.

One down, and a season filled with dozens of crickets and katydids to go. One of the reasons I have been hunting these insects is to collect information for a field guide on the night-singing insects of the Northeast. The illustrator for the guide needs live reference. Photographs suffice, but when possible he likes to make his sketches from the living, moving creatures. As mentioned earlier, hunting for crickets and katydids became my job. Fortunately, it's something I enjoy!

Fens, marshes, and meadows play host to other singing Orthoptera besides the Orchelimum. This is also the domain of the coneheads (genus Neoconocephalus). Named for the conical projection most bear atop their head, the coneheads can present one with another worthy challenge when it comes to finding them. As with the Orchelimum, these insects look like the surrounding monocots: grasses, sedges, and rushes. They tend to rest upside-down on the stems of the plants, their tapered, bladelike outer wings blending with the background. When threatened, they don't jump or fly. Instead they shimmy down the stem to the base of the plant and bury their head in the wild tangle of grass. Once a conehead has taken this position, it's very difficult to locate. I remember finding an Eastern Swordbearer conehead that, upon seeing me, went to the ground in this manner. I watched it do this and saw where it went. When I parted the sedges to fish it out, it was gone. Well, not gone—it was there, but I wasn't seeing it. I had lost it completely, despite knowing exactly where to look. I raked through the grass with my fingers, hoping to cause it to move and give up its location, but to no avail.

Getting the initial location of a conehead requires the same techniques employed in hunting meadow katydids. You can start by homing in on the call, which in this group is often represented by a loud, continuous or broken buzz. Upon seeing you, they will also slip around to the opposite side of the stem, but because they can be larger than the Orchelimum, it's a little less effective. I've found that the easiest way to catch a conehead is to anticipate its escape plan. If you hold your jar or net along the stem below the insect, you can usually goad it down and into it.

A word of caution when handling a conehead: keep your fingers away from its mouth. They have strong mandibles and can give you a good bite. You also want to avoid holding two of them within reach of one another. I made this mistake once when I asked my wife, Betsy, to hold a male and female Long-beaked Conehead for me while I photographed their colorful faces. When Betsy moved the two insects closer together so I could get a photo of them in the same frame, the male conehead grabbed the front leg of the female and began eating it. The leg disappeared like a piece of spaghetti sucked up through a pair of saucy red lips. The more my wife tried to separate them, the more entangled they became. The female conehead, which was now down to five legs, quickly devoured the male's hind tibia, which had been pulled off its body, along with the still-attached femur. Needless to say, this carnage was more than Betsy had agreed to when I asked her to hold the bugs for a photo. I helped her disentangle the two combatants and set them free—in separate corners of the meadow, should one decide to settle old scores.

While the calls of coneheads are not necessarily of a quality the average person would embrace for their listening pleasure, I find most of them to be not unpleasant. The Eastern Swordbearer and Long-beaked Conehead calls have a gentle "shishing" rhythm to them. The Round-tipped Conehead, among the smallest in the group, sounds like a buzzing electric wire. The Nebraska Conehead, whose call has a somewhat angry quality to it, may be more of an acquired taste. Few, however, would sit back, throw on a pair of headphones, and crank up the volume to listen to the song of the Robust Conehead, which has been likened to that of the Dog-day Cicada.

If you want to get into whose songs would fall most kindly on a discerning ear, we'd have to jump to the crickets.

Ground crickets are among the easiest to find. This is despite the fact that they are the tiniest within the Orthoptera group. Ground crickets range in length of one-quarter to half an inch long. They are a variety of dark, earthy shades; amber, ochre, yellow-brown, and black. There are about twenty-five species in North America, with six of those found in my home state of Connecticut. Three of them reside in my own backyard: Allard's, Tinkling, and Carolina Ground Crickets. Striped Ground Crickets, the largest of the group, live just around the corner from me.

We all hear these crickets, and we hear them day and night. The Allard's and Carolina Ground Crickets give a continuous trill from the edges of our yards. The Tinkling Ground

Robust Conehead *(Neoconocephalus robustus).*

Crickets break their trills into a series of "tinks," sounding like tiny bells. The Striped Ground Crickets, which prefer sandy habitats, scratch out a burry "chit . . . chit . . . chit . . ." To find any of them, one need only look down at the ground. These are the tiny dark crickets scampering at your feet when you walk across your lawn. A sweep of the grass with a net is probably the easiest way to catch them. They can also be found beneath the leaf litter and under rocks and logs. One method of catching these insects is to "prospect" for them. Spread a white sheet on the ground and throw handfuls of dead leaves on top of it. Stir it up a bit, and remove the leaves by hand. The crickets should fall to the sheet, where their dark colors stand out against the white background. They can then be scooped up in a jar.

Another technique utilizes a pitfall trap. Bury a jar in the dirt near the area where you hear the crickets calling. Be sure to leave the mouth of the jar just a little bit lower than the surrounding dirt. Sometimes I add a little bait to the trap; fruit and/or pet food works well. Then place some kind of shelter over the jar—leaves, a log, or a flat rock will do—and leave it overnight. This could collect a variety of little critters, including crickets.

The simplest way to scare crickets up from the grass is to "stampede" them. Position your feet so your toes are pointed at ten and two o'clock, and shuffle forward, keeping your feet low to the ground. You'll look like you're doing a Charlie Chaplin imitation. You should be able to notice the crickets hopping out of your way. If you have the opportunity, drive the herd to an open path where they are easier to see and catch.

A few years ago I gathered some friends to hunt for the smallest cricket in North America, the Sphagnum Ground Cricket. These tiny insects live solely in open sphagnum moss bogs. Getting to the bog involved a hike through the woods and then a balancing act atop a very long beaver dam. Upon reaching the edge of the bog, we began to pick up the faint trill of the crickets. It was a soft, high-pitched sound, produced by the blending of many chorusing males. There is an ethereal quality to the song, and when mixed with the croaking of the ravens in the background, the rare and ancient habitat in which they reside becomes all the more magical.

Walking on a quaking bog is an experience in itself, much like walking on a large, soggy trampoline or a waterbed. At no time was the possibility of falling through and sinking beneath the moss far from my thoughts. Visions of those ancient "peat men" dug up in old European bogs kept a constant tendril of caution in my movements. I wondered aloud what future excavators would make of me with my bag of empty jars, butterfly net, digital recorder, and camera. Perhaps I'd be viewed as some hapless shaman, or just another unfortunate bug collector.

Author's friend, Andy Brand, flooding Sphagnum Ground Crickets.

Though the sound of the sphagnum ground crickets was everywhere, we could not see any. In 1904, Canadian entomologist Edmund Murton Walker wrote of a technique one could apply in this instance.[1] If you push down on the moss, it creates a puddle. Any little critters living there would be forced to climb to the surface to avoid becoming submerged. We tried this a number of times, eventually with great success. Our reward was a couple of jars filled with singing crickets, which were kept alive with the moss they lived in and fed upon, and some crumbled dry dog food.

Field crickets look like larger versions of ground crickets. These are the insects that produce the chirping call most of us associate with crickets. The Spring Field Crickets are usually the first of the insects to call after winter has melted away. They call from beneath things—rocks, logs, and any shelter lying flat on the ground. The crickets we hear had overwintered as nymphs. By early spring, they've completed their final molting stage and have grown the wings necessary for song. Spring Field Crickets call day and night. They do take a break in the early morning, but as soon as the sun warms the ground they take to song once again.

There is a second break in the call, and it is one most people fail to notice. It is the brief period of time between the death of the adult Spring Field Crickets and the onset of the Fall Field Cricket breeding season. The adult Spring Field Crickets die by midsummer. Their offspring's nymphs will continue through the season, but will not have their wings to call until the following spring. The nymphs of the Fall Field Crickets develop into adults by mid to late summer. There is an eclipse of sound, sometimes lasting several weeks, as the old are replaced by the new. I've found that I am more likely to notice that eclipse when the calling resumes. It's along the lines of "Oh yeah, it *has* been quiet the last couple of weeks . . ."

It is very easy to find field crickets. They wedge themselves into nooks and crannies of any object or structure lying flat on the ground. I frequently come across them in among the rocks I use as garden borders. My wife made some cement stepping stones for the lawn, and it's a given that there will be a field cricket under any one of them, at times accompanied by Carolina Ground Crickets. Laying out small boards in your yard will also entice the crickets, and other wildlife. I enjoy bringing a few of the field crickets into my studio every season. They call readily, and with a pitch I find pleasing to the ear.

Sand Field Crickets are a handsome representative of the field cricket, or Gryllus genus. They are found along the coast from Connecticut to Florida, growing more common and cosmopolitan you head south. In appearance they're nearly identical to Spring and Fall Field Crickets, but they tend to have straw-yellow tegmina, as opposed to the black to brown tegmina of the other field crickets. Roll over a log on the beach, and if they're there, they will quickly scamper to cover beneath the nearest rock, log, shell, your shoe, or whatever is available. They share that rich chirp of the other two field crickets, but it is slightly lower in pitch. They do hybridize with Fall Field Crickets, so it can sometimes be difficult to distinguish them.

There's another, relatively new cricket in the Southeast that is a distant cousin of the field crickets. The Japanese Burrowing Cricket arrived on this continent in the 1950s and has since spread from New Jersey down to Florida. As their common name suggests, they live in burrows. The call consists of a series of burry chirps, strung closely together. They call from the mouth of their burrows and upon your approach will quickly retreat within. For years I had tried to catch a male but had no success. I'd get right on top of a singing individual, but as soon as I scooped down with my jar, it would retreat as fast as if it had been sucked down the hole.

I was out recording calls in Cedarville State Forest in Maryland one night when the call of a single Japanese Burrowing Cricket dominated my headphones. I circled the car, using the shotgun microphone to find that sweet spot. I went around and around the car, but could not figure out from which direction it was coming. This was odd, since I can always get, at the very least, a rough idea of the source of a call. I walked away from the car and scanned the area some more. The sound grew stronger when the microphone was pointing at the car. I got down on my belly and searched beneath the car with my flashlight. Nothing was there. "It's in the car!" I said. I opened the door and started pulling out my stacks of jars, books, and clothes—everything I had packed for a week's worth of hunting insects. No cricket—but the chirping was still going as strong as ever. I put everything back in the car. This was a real puzzler. I got back down on the ground and crawled under the car, where the call seemed loudest. It reverberated beneath the undercarriage. It sounded so close! Was it caught up somewhere within the greasy chambers in the underside of the vehicle? If so, it obviously didn't mind being there, as evidenced by its incessant singing. I searched with no results and slid back out. Giving one more flashlight sweep of the road beneath the car, I noticed nothing but a single dead leaf. I decided to take a closer look and crawled back again. The call stopped. I looked under the leaf and there it was. A lone male Japanese Burrowing Cricket peeked out from the edge.

I've read stories about the exhilaration deer hunters feel when their prey moves into range. Sometimes their bodies begin to shake in excitement. At that moment, I experienced that very same phenomenon. I actually froze for a moment, afraid that the wrong movement would make the cricket bolt from its hiding place. Do I go for my camera and take its picture? Do I try to catch it right now with my bare hands? Do I get a jar

Japanese Burrowing Cricket *(Velarifictorus micado)* under leaf beneath car.

and take the chance it might be gone when I crawl back under the car?

I backed out from beneath the car, got my camera and a jar, and managed to nab this elusive singer. Having finally accomplished something I had failed at for quite some time, I imagine people in the houses I drove past must have received an unexplainable warm feeling from the rush of elation emanating from my being.

There is another cricket in the United States that lives in burrows. It's the Northern Mole Cricket, and it is found from Maine to Florida. They spend most of their lives within their tunnels, the females leaving only to fly to a serenading male. The males sit and call near the edge of their hole, using it as an amplifier much in the way the acoustic horn worked for those old Victrola record players. I remember the very first time I heard these insects. I had no idea what I was listening to. I tried to place it in some kind of category, and even entertained "frog," although I knew of no frog that sounded like that. The call is a steady "dirt-dirt-dirt-dirt . . ." It had a somewhat croaking quality to it. As it turned out, the crickets were telling me where to look for them—in the dirt!

I'd learned of a place where a colony of these crickets was likely to be found, and decided to attempt to catch one. I did not know for a fact they would be there, but their presence was suggested by the raised ridges of their tunnels at the edge of Horse Pond in Madison, Connecticut. Horse Pond consists of the pond and a thin line of trees surrounding it. Those in turn are surrounded by houses. It's a little oasis in a suburban setting.

I decided to try a linear pitfall trap I'd read about in the publication *Florida Entomologist.* To make this contraption, you have to cut a wide slit, lengthwise, in a long, hollow tube—I used PVC pipe. One end is blocked off with duct tape and the other leads into a container. For the container, I used a cleaned-out bleach bottle. You bury the pipe horizontally so the slits on top are just a little below the soil surface. As the mole crickets travel through the dirt, some are supposed to pass over the tube and fall into the slits. They then follow the length of the pipe and fall into the container.

I buried the trap in the mud on the edge of the pond. It was in an area where the displaced dirt of several tunnels suggested mole cricket activity. When I returned the next morning and emptied out the bleach bottle, I was happy to see about a half a dozen mole crickets crawling about. However, all I managed to catch were nymphs. I wanted to catch an adult so I could hear it call. I removed the trap and decided to just dig with a trowel. Had I known how well that would have worked, I would have saved myself the trouble of building and implementing a real impressive-looking trap, but one that was ultimately unnecessary

Mole cricket trap.

for my purpose. About every fifth hole I dug produced a great big fat juicy adult mole cricket! One only needs to dig about a foot, because most of them stick fairly close to the surface, where they feed on plant roots, worms, and grubs.

I brought a couple of crickets home, hoping to enjoy their call for a while in my studio. I packed them up with the soil I found them in and put them in my dark closet. That night they sung like nightingales. Nightingales with sore throats, but to my ears it was pure sweetness.

Mole, field, and ground crickets are all creatures of the soil and dead leaves. There's another group of crickets that set their sights a little higher; the tree crickets. While the name "tree cricket" may suggest they offer yet another challenge when it comes to hunting down an individual, fortunately they are not always up in trees. This group, which includes the Neoxabea and Oecanthus genera, can often be found in lower locales, and yet always off the ground.

As with the field crickets, tree crickets can produce a call pleasant to the human ear. They tend to trill, either continuously or broken into pulsing sections. Three species in my area, Two-spotted, Snowy, and Davis's Tree Crickets, inhabit various levels of the deciduous leaf canopy of trees and can certainly be out of reach. You can hear them and identify them by call, but catching them is a challenge. However, many times I've come across them in low-lying fauna at the edge of the woods. Sometimes they are blown from their perches in storms, and a search around the surrounding vegetation the next morning may turn them up. The three deciduous tree toppers may also be encountered at porch lights, or at any other light shining in the night.

Pine Tree Crickets are found at various levels in trees, but on conifers. Their rich continuous trill can be heard day and night from midsummer through autumn. The rest of the tree cricket clan frequent the lower herbaceous and woody plants in fields and the lower leaves of trees. The Pine Tree Crickets, when calling from lower boughs, can be found by using the same triangulation method—hands cupped behind the ears— as mentioned in my hunt for the meadow katydids. Fortunately they can be more forgiving of your presence and may continue to call while you are just inches away. The problem lies in the fact that their call will bounce off of the surrounding branches, effectively masking their location. I've spent a lot of time over the years circling conifers trying to locate an individual Pine Tree Cricket that sounded as if it was right in front of me. As soon as you think you've locked on to that sweet spot of sound, it seems to move to a different location. It's almost as if your movement is pushing the sound just out of reach. The reward of the hunt, however, is actually getting to see one of these spectacular insects. They are an attractive mix of deep greens and browns, looking just like the pine needles that surround them. To my ear, they have the most pleasant call, too. A copse of pines filled with singing Pine Tree Crickets leaves one with the impression that the trees themselves are singing.

Rounding out the numbers of the rest of the tree crickets in my area are the Four-spotted, Black-horned, Narrow-winged, and Broad-winged Tree Crickets. They can be found day and night by following their calls to the source. A goldenrod-filled meadow is a great place to look for these insects during the day. Only the males will call; the females can often be found in their vicinity.

Narrow-winged Tree Crickets are probably the most encountered species in the northern and eastern United States. Their call is a steady, pulsing trill, much like a ringing telephone. I often find these in the lower leaves of deciduous trees and on woody and herbaceous plants in open meadows. They resemble slender little stems, their pale green coloration broken only

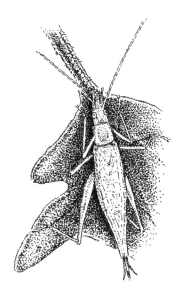

Four-spotted Tree Cricket (*Oecanthus quadripunctatus*).

by an orange diamond on top of the head. A good way to find Narrow-winged Tree Crickets at night is to walk under the lower leaves of the tree canopy and search with a flashlight. Many rest on the ventral surface of the leaf, which is most likely a habit they evolved to elude avian predators. In fact, that pale coloration is a closer match to the bottom of a leaf than it is to the upper surface. The singing males will often chew a rounded hole through the leaf, poke its head and wings through to the top, and trill away. The wings are held up at about a forty-five-degree angle from the body, forming a blurry, transparent heart. You may be fortunate enough to encounter a female mounted on its back and feeding on the nuptial treat it secretes from under the base of the wings.

The Broad-winged Tree Cricket is the loudest-calling of this group, which can likely be attributed to the fact it has the widest wings. A greater surface area can only increase the resonance and volume of the call as those wings are rubbed together. I tend to come across them in raspberry plants. The call is a highly resonant and continuous trill. These crickets are as much a treat for the eyes as they are the ears. This is the only species with a vibrant raspberry-red head and antennae bases.

The Black-horned and Four-spotted Tree Crickets, and in the South, the Fast Calling Tree Crickets, are found in weedy meadows in the flowers and shrubs. Goldenrod seems to be popular with them. They call on the sunniest of days and the darkest of nights.

The most famous tree cricket is more of a chirper than a triller. This would be the Snowy Tree Cricket, the species people have long been able to count on to tell them the outdoor temperature. I wrote more about this species in Chapter 7.

When I lead insect walks, I enjoy pointing out the tree crickets to people who have been hearing them their whole life but never imagined what they look like. These crickets are such delicate looking creatures, and yet their calls can dominate the sounds of the night. The real crowd pleasers, though, are the bush and round-headed katydids. Both genera share the subfamily Phaneropterinae, or False Katydids. Their popularity among naturalists is due largely to their strikingly similar appearance to leaves. Most within these two genera, the Scudderia and Amblycorypha, are very accessible, as they inhabit the lower vegetation in fields and forest edges. They are easily approached, easily captured, and easily handled. I suppose they look so much like leaves that they don't feel the need to be skittish.

The Scudderia, or bush katydids, produce a scratchy, staccato sound. The call is fairly soft, compared with that of their noisier neighbors, but once you develop an ear for them, you won't miss them. I have found that the best way to catch one is to simply scan the low deciduous plants with a flashlight. They are often on the cherry and sassafras saplings they feed upon. Again, I find that the antennae tend to jump out at me, because they look less like a leaf than the rest of the insect, which, not coincidentally, will be sitting on a leaf.

John D. Spooner has described how female bush katydids click in response to calling males.[2] You can call males to you by imitating the female's response. Spooner suggested tapping the tip of a knife against the curved glass edge of a bottle, 0.1 to 1.5 seconds after the male's call. This also works for the round-headed and angle-winged katydids. Patience is often necessary, as the males rarely come *rushing* to the clicking female. I tried this recently with some positive results for Curve-tailed Bush Katydids. I've also had duets with Greater Angle-wings by tapping two pebbles together.

Should you desire to bring one home, you can either pick it up with your fingers or scoop it up in a jar. Many can also be coaxed to an outstretched hand. Most insects can't tell the difference between a branch and a hand, as long as that hand is held still.

The bush katydids are most easily separated in the field by the different shapes at the tip of the male's abdomen. Fortunately for those of us who care to make the distinction between one species and another, those appendages are fairly easy to see with the naked eye. To properly handle a katydid in order to get a better look at its "nethers," you will need to pinch the hind "knees" together between thumb and forefinger. It helps to grab a little piece of the upper wing, too, to keep it still. Never grab a katydid by the leg, as it will snap right off.

The Amblycorypha, or round-headed katydids, are handled in the same manner, although they are easier to tell apart based on the shape of their wings. In my backyard I have all three New England species: Oblong-winged Katydid, Rattler Round-winged Katydid, and Carinate Katydid. The Oblong-wing and Carinate Katydids both give a loud "skritchy" call, the Carinate's is steadier and closer together. The Rattler Round-winged Katydid gives a sputtery call that sounds to me like someone trying to start a tiny, flooded, two-cycle engine. These are the tamest of all the katydids, and like many of the bush katydids, they can often be easily coaxed to your waiting palm. If you want to have some fun, offer one a slice of apple. Once they are enticed to step aboard, they will eat with such fervor, they will have no idea they're being held by a giant predator. What's interesting to me is that an insect that has never eaten an apple in its life will so easily take to the foreign food being offered. I can't help but think what a once-in-a-lifetime treat this is for a gentle friend.

There's another genus in the False Katydid subfamily that is also present in good numbers throughout their range. The Microcentrum, or angle-wings, live in the higher reaches of deciduous trees. The name Microcentrum means "small

Rattler Round-winged Katydid *(Amblycorypha rotundifolia)* enjoying an apple treat.

point" and refers to the wingtips, which are sharply tapered. These are also incredible leaf mimics, and are represented by two species in the eastern United States: the Greater and Lesser Angle-wing.

The Greater Angle-wing is the northern relative of the Lesser Angle-wing. One of its calls consists of a series of ticks, much like the sound of two pebbles being tapped together, or a Geiger counter (a new nickname for this insect). The Lesser Angle-wing, found south of New York, gives a lisping rattle, sounding like a rushed version of the call of the Common True Katydid. Your best chance of finding either of these lies in a search of low-growing trees. The angle-wings inhabit the upper stories of trees, but if those trees are fairly short, then it's sometimes possible to find them without a death-defying climb. Fruit orchards, where the trees have been cultivated to attain heights more in line with ease of harvesting, can provide an excellent opportunity to catch one of these katydids.

The same can be said for the noisiest katydid of all, the Common True Katydid. This is the insect whose call lent the name "katydid" to this group of insects. They sing from the tops of the trees, their "tch-tch-tch . . . tch-tch-tch . . ." being likened to the mnemonic "kay-tee-did . . . kay-tee-did." Few people get to see this insect, whose call dominates the night from Vermont to Florida. As frequently as I am out in the field, day and night, I've seen only five individuals in my lifetime. One of them was as a child growing up in Oceanside, New York. The katydid was on the screen door to our house, where it was probably attracted to the light. When I picked it up, it gave its alarm call—a loud, single-noted "scratch." I was so startled, I dropped the insect, proving what an effective defensive technique its call is!

Every other time I've found these katydids, it was by pure chance. In each case the katydid was either in the low vegetation or on my house. I found one healthy male Common True Katydid while forging across a stream. It was in a tussock sedge, and the only reason I noticed it was because the oblong shape contrasted with the narrow blades of the sedge. Had it been a conehead or meadow katydid, which evolved to blend with this kind of plant, I would have walked right by it.

If you want to see a Common True Katydid for yourself, you can search the younger, lower trees or orchard trees, or just count on chance. I suppose you could bring them down to you with the aid of a chainsaw. I'm only partially joking about that. If you do have to cut down trees for some reason, you may have the opportunity to search the fallen crowns for any of the upper-canopy insects. While I'm sure many will "abandon ship" as the tree is falling, there's a chance that some will ride it down to the ground.

Finding crickets and katydids requires a little bit of patience, a little bit of skill, and a little bit of luck. Patience you're born with, or without. Skill comes with practice. Luck comes when you put yourself in places where these creatures live. It's the kind of luck you make, or earn. It is very common to come across one insect while looking for another. In fact, some of my most exciting finds have come about as a result of stumbling across a bug that did not belong where it was. Insects don't read field guides, so they can be forgiven for these transgressions.

nine

The Bug People

Putting Everything in Its Right Place

As a seventeenth-century buccaneer sailing aboard the notorious Admiral Henry Morgan's pirate ship, John Esquemeling witnessed many a bloody battle. He was privy to the burning of Panama and the ravaging of coastal Cuba. Granted, he didn't do much of the burning and ravaging, but he was on hand, watching. The pirate was a keen observer, with a curious mind and a gift for the ability to describe what he saw.

Esquemeling began his path in history as a clerk on a French West India Company ship sailing to the island of Tortuga to sell supplies to French colonists. Although the colonists, many of whom were pirates, welcomed these wares, they had different ideas when it came to payment for goods received. Getting a buccaneer to part with his booty was no easy task. The West India Company gave up on trying to get the full amount owed to them and, to reduce their losses, sold every item on the ship that was not needed to get them back home. One of those *items* was a young clerk named John Esquemeling. The new slave found himself bound, in his words, "to the most cruel tyrant and perfidious man that was ever born of a woman." As a result of his mistreatment, he became very ill, and to prevent a total loss of his investment, Esquemeling's owner sold him to an

Cover illustration for Esquemeling's *The Buccaneers of America*.
Artist: George Alfred Williams, for *The Buccaneers of America*,
by John Esquemeling, illustrated and translated in 1914.

island doctor for seventy pieces of eight. Selling a sick slave in this manner was a fairly common practice, not unlike a sick horse being sold to a veterinarian. The thought was that if the doctor could nurse the patient back to health, he had procured a slave, or horse, for a good price.

Esquemeling recovered, and the doctor turned out to be a more decent man. He freed Esquemeling under the promise that he would some day pay the doctor back 100 pieces of eight so the doctor would have a profit on his investment. The newly freed man, "naked and destitute of all human necessaries, nor knowing how to get my living," made a decision: "I determined to enter into the wicked order of the Pirates, or Robbers at Sea."

In his descriptions of pirate life in *The Buccaneers of America,*
the account he later published about his exploits, we learn not
only about the men, their ships, and their conquests, but also
a bit of the natural history of the New World.[1] Esquemeling
wrote detailed accounts of the flora and fauna he came across,
including insects. While in Hispaniola (the island of Haiti and
the Dominican Republic) he wrote about the mosquitoes and
flies that plagued him night and day, and how the local hunters had to smear themselves with "hog grease" to keep them at
bay. He had a more pleasant experience with fireflies, though:

> These are very like such as we have in Europe, unless
> that they are somewhat bigger and longer than ours.
> They have two little specks on their heads, which by
> night give so much light that three or four of those
> animals, being together upon a tree, it is not discern-
> ible at a distance from a bright shining fire. I had on a
> certain time at once three of these conchinillas in my
> cottage, which there continued until past midnight,
> shining so brightly that without any other light I
> could easily read in any book, although of never so
> small a print. (29)

He was also quite impressed with the crickets: "There be
also in Hispaniola an excessive number of grillones or crickets.
These are of an extraordinary magnitude, if compared to ours,
and so full of noise that they are ready to burst themselves with
singing, if any person comes near them" (29).

Esquemeling stayed on with Admiral Morgan until 1672
and published his account of his adventures after he returned
to Amsterdam. While the book is mostly about the pirates, it
contains some of the very first descriptions of insects in the
New World.

With regard to the crickets, it's hard to say if he was referring to the chorus of a single species or the mélange of several.
When I try to listen for the burst of song heard through the
ears of the long-gone buccaneer, I hear a ringing, pulsing trill.
It is what I remember from my own visits to that part of the
world. In listening through the words Esquemeling wrote of
crickets that died three and a half centuries ago, I am actually hearing in my mind's ear the crickets I heard when I last
visited the area. Those singers are now as dead as those from
the buccaneer's time, but I imagine that when they were alive,
they sounded no different from their ancestors.

What confused me initially about his short mention of the
crickets was how they would get louder "if any person comes
near them." Normally when one approaches a calling Ensiferan,
it stops singing when it notices that presence. What I believe
Esquemeling is describing is how, as you get close to a single
member of a great chorus of insects, your ear can pick out its
call from that chorus. Once you home in on that individual
singer, it seems quite distinctive in relation to the others calling
from farther away. This phenomenon often occurs when I am
listening to music. Once I pick up a single instrument in the
ensemble, my ears won't let go of it. That one little tidbit added
by Esquemeling traveled over three centuries and thousands of
miles to bring us together in the field, for a moment.

I have always marveled at how the written word can connect
one to distant lives. It is a gift the writer gives the reader, an
insight into his or her own thoughts and experiences that live
on long after that writer is gone. I was just listening to crickets with one of Captain Morgan's pirates and can almost see
his face and read his body language as he approaches them,
amazed at the great volume of sound they produce. What's
missing from the picture is what he is actually seeing. I can
visualize the lush tropical plants surrounding him. I can feel

the warm tropical air. I can hear the sounds. However, when my mind's eye looks to the ground, or to the leaf from which the cricket is calling, I see something akin to that "Site Under Construction" banner we read when calling up a dead page on the Internet. There are just too many possibilities. It could be a tree cricket. It could be a bush cricket. It could be a ground or field cricket. Each of those Orthoptera categories offers a wide range of individuals from which to choose.

This process of trying to fill in the blanks follows a thought process that takes us back to the very origin of word formation. I believe it is reasonably safe to assume that our language arose by assigning names to things important to us. These could be things that were safe to eat or things that made us sick. Over time our brains developed the ability to group these things into categories. For example, I can envision an early human pointing (some believe gestures came before speech) to a number of plants and thinking, "Good to eat. Good to eat. Bad to eat. Good to eat. Bad to eat." Those plants would be categorized as "Good" and "Bad," based upon learned experiences. This not only served to whittle down the possibilities within the individual's brain, but aided in the sharing of that information with others. It formed a key component in communication.

To give a thing a word is to give it a certain universality. To define that word, whether in the mind or on paper, is to lock it in that universality. To put it in a category is give it both a commonality and exclusivity. This has long been the task of taxonomists and systemologists, who seek out the connections in the natural world. They are creators of pyramids, the peak of the pyramid being that which most things share and the base being filled with that which is unique. It could be said that Adam was the first taxonomist, as one of his first jobs on this new planet was to give names to all the creatures.

Over the years I have become obsessed with the pyramid of

singing insects. Having enjoyed the songs of so many crickets and katydids in the field, I wanted to learn more about them. To give another music reference, it's like hearing a song and wanting to know the name of the person who sang it. When you want to hear more of that style of music, you can listen to other groups within that category. I'm not sure what is at the very peak of the music pyramid; perhaps it's just "sound," or "air vibration."

In learning about the crickets and katydids that create the sound I so enjoy, I've come across the names of a number of interesting people, Esquemeling being one example. These are people who came before me on this planet and gathered and shared the information that has made this a fruitful pursuit. Many of them have their name attached to a particular insect I've come across. When I hear the Allard's Ground Crickets in my yard, I think of Harry A. Allard, born on a farm in Massachusetts in 1880. *Allonemobius allardi* are a common and widespread species. Their rich, continuous trill is a welcome background to my summer evenings outdoors. Allard's main claim to fame in the world of science is his co-discovery of photoperiodism in plants, but he was making some waves, sound waves in particular, with his work on singing insects. At the turn of the twentieth century, Allard described the three distinctly different calls and ecology of what were then thought to be different forms of the same ground cricket: *Allonemobius fasciatus* (Striped Ground Cricket). He was the first to note their biological differences in a clear and accurate manner. In 1959 entomologists Richard Alexander and Edward Thomas separated these forms as full species and honored Allard's work by naming one of them after him. This was four years before his death, and it's nice to imagine old Harry sitting out on his porch with his grandchildren, saying, "Listen . . . Hear my crickets?"

Then there's the Caudell's Conehead, *Neoconocephalus caudellianus,* a southeastern species of katydid that inhabits the grasses and sedges of freshwater habitats. In 1905 W. T. Davis named it for Andrew Nelson Caudell, an entomologist born in Indianapolis in 1872. Caudell was a colleague of Allard's, and in 1930 the two had worked on the manual *Orthoptera of the District of Columbia, Maryland, and Virginia.* It was never published, but Caudell's work in Orthoptera while custodian of this group of insects at the U.S. National Museum and entomologist for the Division of Insects of the U.S. Department of Agriculture deserved recognition.

There is a portrait of Andrew Nelson Caudell in the *Proceedings of the Entomological Society of Washington* (1936). He is looking at the camera; pleasant smile on his face; wearing a dark suit and tie. His hair is short on the sides but glides up into a swooping wave on the top of his head. I can't help but associate that

Andrew Nelson Caudell.

with the cone of the conehead, which bears his name. It is very nearly the same exact shape! Coincidence? Likely, but if I had a conehead named for me, sure, I'd have fun with my portrait.

Davis, the man who named that conehead after Caudell, has an insect named for him, the Davis's Tree Cricket. This is a somewhat unusual situation, however, in that the common name of this insect is that of the person who originally described it. The namer rarely makes *himself* the honoree. In most cases where a person is honored by having his name attached to a species, that species name is a Latinized form of the name of that honoree. For example, Caudell's name becomes *caudellianus* in the species name. The scientific name of Davis's Tree Cricket is *Oecanthus exclamationis*. That's the name Davis gave it in 1907. "Exclamationis" makes reference to the exclamation point pattern on the base of the antennae. References to this species in *Guide to the Insects of Connecticut* (1911) give it no common name, although the other tree crickets possess them.[2] As far as I've been able to discover, it wasn't until 1920, in two separate books on Orthoptera, by Samuel Morse and Albert Pierce, that the cricket was first given the common name—Davis's Tree Cricket. I suppose it flowed better than "Exclamation Point Tree Cricket."

William Thompson Davis was born in Staten Island, New York, in 1862. He was a well-rounded naturalist but held a special interest in cicadas. He had named and described 100 of the nearly 170 species known to North America. Davis was first to discover *his* tree cricket in New Jersey where he was "arrested by the character of its song." Bentley Fulton, author of *The Tree Crickets of New York,* wrote of the call: "In quality it most resembles the distant singing of the common toad."[3] Fulton, by the way, was also honored with the naming of a tree cricket after him. In 1962 Thomas J. Walker of the University of Florida carved it from the species *Oecanthus niveus,* where it had been incorrectly placed, and named it *Oecanthus fultoni*. However, the common name, Snowy Tree Cricket, does not reflect this honorific. The common name had been borne by this insect for many years prior.

The same thing happened to Henri Louis Frédéric de Saussure, whose *Miogryllus saussurei* is known as the Eastern Striped Cricket, not as "Saussure's Cricket." I suppose a choice must be made between assigning a name that tells you something about the organism, and honoring someone who has contributed to the knowledge of that insect or group of insects.

There are many scientists no longer among the living who contributed to the field of Orthoptera study. The names Johann Christian Fabricus, Lawrence Bruner, Carl Stal, E. M. Walker, and Philip Reese Uhler appear frequently in the literature. Some of those people go back the mid-eighteenth century. When it came to taxonomy, the naming of things and putting them into categories, they did have something to work with. This was due in part to the work of one of Plato's star students in the fourth century. Prior to that, in our early-developing grasp of the natural world, the hodgepodge of plants and animals about which we were learning, lacked a system of order.

Aristotle, son of a physician to the king of Macedonia, was encouraged at an early age to pursue scientific studies. His father died when Aristotle was a boy, but his caretaker continued to foster his intellect, and when Aristotle turned seventeen, he was sent to Plato's Academy. For twenty years he lived in Athens and studied under the great philosopher. Following Plato's death, Aristotle, who had risen in stature within the intellectual and royal circles, spent several years traveling, learning, and teaching. At one point he became the private tutor to the son of a Macedonian king. History refers to that son as Alexander the Great. The scholar also took some time to study botany and zoology on the Island of Lesbos.

In 335 BC, Aristotle created his own school, called the Lyceum, located outside the walls of Athens. Around this time he wrote his treatise *Historia Animalium* (History of Animals), which was one of the earliest attempts to create a zoological system of order.

History of Animals is an interesting read. It was translated by D'Arcy Wentworth Thompson in 1910, and is fairly easy to get hold of.[4] Nearly all of the information it contains was gathered by thousands of men sent scouring the Grecian countryside, at the king's command, for all manner of living things.

Aristotle filled in the blanks in his personal research with what was the common knowledge at the time. He was in no way the first to observe the natural world, but he was one of the first to leave behind his observations.

His system began with the division of living organisms into two major groups: land animals and aquatic animals. These were further broken down by physical characteristics, how they lived, and where they lived. He also differentiated between what he believed were animals with blood and animals without blood. The insects were arranged within the system of "animals devoid of blood." While insects do not have blood in the sense that mammals have blood, they do have a bloodlike fluid (hemolymph) pumped through the body. That fluid is moved through an open system, as opposed to our circulatory system, which carries blood through veins and arteries.

The bloodless animal category was broken up into four groups: Molluscs (soft on the outside, hard on the inside— an octopus is used as an example); Malacostraca (hard on the outside, soft on the inside—their shells crush rather than shatter; crabs fell into that category); Testaceans (snails, sea urchins, and scallops—which he believed could fly); and Entomos (includes insects and arachnids).

"Entomos" means "having a notch or cut" and refers to the

segmentation of insect bodies. The word *insect* means about the same thing—"cut into." Aristotle obviously picked up on the fact that insects have three main body parts: "Insects have three parts common to them all; the head, the trunk containing the stomach, and a third part in betwixt these two, corresponding to what in other creatures embrace chest and back."

The next paragraph creates a picture of the nature of the research he put into some of his entries. I suppose it describes a form of vivisection:

> All insects when cut in two continue to live, excepting such as are naturally cold by nature, or such as from their minute size chill rapidly; though, by the way, wasps notwithstanding their small size continue living after severance. In conjunction with the middle portion either the head or the stomach can live, but the head cannot live by itself.

Did he actually observe this, or was he passing on what he had read or heard? It could be a little of both. Many insects continue to twitch for a while after they're killed. This could be what he was referring to. Upon reading this account in his treatise, my mind formed an image of a man in a toga, sandals on his feet, cutting up various insects that were brought to him. He watches, brow furrowed, as the bodies continue to twitch and then scribbles some notes on a clay tablet or parchment.

Sculptors of the day have given us an idea of what Aristotle looked like. In 2006 Greek archaeologists unearthed a bust believed to be the best-preserved likeness of the philosopher. That sculpture shows him to have had a robust face, heavily bearded, and with the hardy features one may equate with those of an old New England sea captain. His legs, however, have been compared to stalks of parsley. Whenever I see the faces of antiquity, created in antiquity, I try to imagine them donning a fedora or baseball cap. This helps me place them more in context with a real person. Gone is that mythical veil that can shroud the image and accomplishments of someone virtually canonized by centuries of historical writing. In a way it increases my appreciation of those accomplishments, because they were achieved by someone who resembles a person I might run into at the supermarket.

Of insect song, Aristotle mostly writes of cicadas. He makes a brief mention of how grasshoppers produce noise by "rubbing or reverberating with their long hind-legs."

Most interesting is a section on insect reproduction:

> Other insects are not derived from living parentage, but are generated spontaneously: some out of dew falling on leaves, ordinarily in spring-time, but not seldom in winter when there has been a stretch of fair

A dapper Aristotle.

weather and southerly winds; others grow in decaying mud or dung; others in timber, green or dry; some in the hair of animals; some in the flesh of animals; some in excrements: and some from excrement after it has been voided, and some from excrement yet within the living animal. (139)

Spontaneous generation in insects was taken as fact for another *nineteen centuries* following Aristotle's work! It was finally refuted by Francesco Redi in 1668 in a well-known experiment proving that fly maggots come from flies, not meat.

Aristotle further divided the insects into more subsets based upon a number of physical features, including mouthparts and wings. For instance, he distinguishes between insects that have "teeth" and chew things and insects that have only a tongue and take in liquid nourishment. He also separates the Pterota, or winged insects, from the Aptera, insects without wings.

Aristotle spent his last year on this planet in fear for his life, as, upon the death of Alexander, the new ruling forces in Athens charged him with impiety for not honoring their gods. He moved to his mother's estate in Chalcis, located on an island east of Athens. He successfully escaped public execution but wallowed in boredom, exclaiming to a friend that the more he was alone, the fonder he became of myths. His longtime battle with digestive problems eventually ended with his death at the age of sixty-two.

While Aristotle got the ball rolling in figuring out which animal was related to what animal, that ball had a very long way to roll before any great new advances were made. Redi's discovery of sexual reproduction in insects was substantial, but much of the rest of Aristotle's work saw little tweaking until 1735. This was the year that Carolus Linnaeus published his first edition of *Systema Naturae*.

Linnaeus was born in 1707 in a small farmhouse in Stenbrohult, Sweden. His father, Nils, was a pastor and an avid gardener, but young Carolus, while pious throughout his life, chose a path exploring the physical rather than the metaphysical. His father's gardens had more than a little influence on his son's interest in nature. Carolus spent much time among his father's plants, tending them and learning their names. At the age of thirty-eight Linnaeus wrote, "This garden inflamed my mind from infancy onwards with an unquenchable love of plants."[5] Even the surname *Linnaeus* has botanical roots. It was chosen by his father, a proper surname being a prerequisite to entering a university at the time. Prior to that, the family surname was Ingemarsson, meaning "Son of Ingemar," Ingemar being Nils's father. Nils chose a name to honor a tree that grew on his farm, the linden, or "linn," as it was known as in that region.

There were few career options available in those days for those with an interest in the natural sciences. Physician was about as close as one could get to making a living in the sciences, so in 1727 Linnaeus set off to Lund University to study medicine. This afforded him the opportunity to study botany, a field he had been driven to pursue since childhood.

Physicians of the time needed to be proficient at creating and procuring the medicine they administered, and much of that medicine came from plants. The curriculum at Lund focused more on the divine than on science, so after a year Linnaeus transferred to the University of Uppsala, considered to be the most prestigious institute of learning in Sweden. He arrived dirt-poor, but with a hunger for knowledge in botany. In Uppsala he attracted the attention of some of the leading botanical scientists of the day and was offered room and board, and assistance in obtaining a grant to collect plants in Lapland. Little was known of "Lappland" at that time, in fact, it was believed to be a place where lemmings fell from the clouds like rain. Linnaeus braved many hardships (and raining lemmings) in the region, and his discoveries and observations so impressed the government that he was sent on a new expedition to Dalecarlia in central Sweden. Here he discovered his most prized flower, Sara Lisa Moraea, whom he eventually married.

Upon his return, Linnaeus traveled to Harderwijk, Holland, and received his doctorate in medicine in 1735. That same year he published his first edition of *Systema Naturae*. This edition was only eleven pages long, but within those scant pages he revealed his new classification for the three kingdoms in nature: animal kingdom, plant kingdom, and the kingdom of stones. The first edition dealt with plant species, the classifications of which were based upon the arrangement of their pistils and stamens. Over time, however, all manner of living things were included. He published a total of thirteen editions, the last in 1770 containing 1,300 pages. His plan was to replace the old system, which used descriptive phrases, sentences, and paragraphs to delineate an organism, with a system that referred to organisms using just two words: those for the genus and the species. This is the binomial system we use today. Linnaeus was not the first to use binomials. There were many plants and animals known by two names since the time of Aristotle and before. What Linnaeus did was create a system that was consistently binomial.

It is the tenth edition, published in 1758, that zoologists have accepted as the starting point of our current binomial system of nomenclature (it's also the edition where whales went from fish to mammal). Linnaeus founded his insect system upon the absence or presence, and the characters, of the organs of flight. This was similar to Aristotle's alary (pertaining to wings) system. In the 260 years since Linnaeus's tenth edition, there have been changes to the system. DNA research has added to the mix another element to be considered besides morphological and sexual features. In addition, many living things now sport trinomials, where the third name indicates subspecies.

The interesting thing about the origin of the binomial system is that it came about as a way to save time and money, as opposed to springing from a perceived need to create order from natural chaos. As Linnaeus added new editions to his original work, the numbers of included species grew. So did the number of contributors, who, in the tradition of the time, would treasure the memories of their finds with a brief description following the name of the organism. By the ninth edition, the species names and descriptions were cluttering the pages. This made it more difficult to remember the names of the growing number of biota treated. It also added to the volume of pages, which increased the printing costs. Add to that the author's compulsion for editorial pithiness, and the stage is set for the creation of a system born of clerical necessity. Linnaeus simply shortened their descriptions by using a generic name (genus), followed by the first or most descriptive adjective contained in the sentence that followed (species). He liked the way that worked, and in his tenth edition of *Systema Naturae* that binomial system of nomenclature became the go-to method of naming species.

Linnaeus's work was still rooted in the same Greek logic that framed the work of Aristotle. In the *Organon,* a collection of his thoughts on logic, Aristotle includes his theory called "The Categories." This collection breaks down "all there is" into ten categories, including such concepts as time, place, substance, and state. What Linnaeus did was take that concept of categorization and apply it to nature, using as few words as possible.

Although the Swedish naturalist chose not to follow in his father's footsteps as a man of the clergy, his understanding of the natural order of the world did not conflict with his beliefs in a higher power. In fact, he had been fond of saying, *"Deus creavit, Linnaeus disposuit,"* meaning "God created, Linnaeus organized."

Carolus Linnaeus lived a fairly long and highly productive life. He died at the age of seventy-one. He had described nearly two thousand insects, most from Europe, but hundreds from America, which were collected and sent to him by friends and students. His best-known insect would be familiar to anyone who has ever been in a pet store: the House Cricket, or *Acheta domesticus*. I wrote more about this species in Chapter 7.

In 1773 one of Linnaeus's students pinned a name on one of my favorite insects, *Amblycorypha oblongifolia,* or Oblong-winged Katydid. That student was Charles De Geer, and his description of the species likely came from dead specimens sent to him from America.

But, one may ask, how would we know that Charles De Geer was the person to first describe this species? That information becomes imbedded within the name of that species. While we now use the system of binomials (and sometimes trinomials) in the scientific literature, there is an additional word tacked onto the genus and species name: the last name of the person who first described it. That name is often followed by the year in which it was described. Sometimes new information leads the taxonomists to move a species to a new genus. The original describer's name still remains attached to that species, but it is put in parentheses.

Amblycorypha oblongifolia will appear in much of the literature covering this species as *"Amblycorypha oblongifolia* De Geer 1773." The scientific name is shown in italics. Many refer to a scientific name as the Latin name, by the way, but while Latin is most often used in this context, it is not always the only language used. Many scientific names contain Greek words or a smattering of words with other origins. The original katydid

De Geer used to describe this species is still in existence and is housed at the Swedish Museum of Natural History.

So Charles De Geer's name will be forever linked to an insect he would only have seen dead. Again, this is not unusual, but fortunately for most of us, we get to appreciate his bug alive and kicking.

The Oblong-winged Katydids are found throughout the central and eastern part of the United States. They inhabit the lower stories of deciduous trees and herbaceous vegetation in meadows. While their call is not the most melodious of the katydids (few katydids carry much of a melody), the sound they produce, to my ear at least, is not unpleasant. It's a short, scratchy "skritch-it" or "ski-deet." What I like about this species is a quality present in most of the Amblycorypha (round-headed katydids)—their demeanor. Oblong-wings move like cats. This is most evident when they know they're being watched. They will rock slowly back and forth, mimicking a leaf in the breeze as they cautiously place one foot in front of the other. The leg movement brings to mind that of a cat that has just spotted prey. They are also quite fastidious, often observed running a tarsus or antenna through their mouths.

I was watching an Oblong-wing one evening as it chewed away at a sassafras leaf. Its mouth was silently and smoothly shearing a hole in the middle. I thought about Charles De Geer and knew he would have been staring at that mouth.

Again, this action was one De Geer had never witnessed himself; nor had he heard that insect's strident call. It's possible that, had he enough specimens to play with, he attempted to make one speak from beyond the grave by rubbing the tegmina's file and scraper together. I've tried this. It makes a noise, but nothing like that of the living bug. As with Linnaeus and most of the European natural scientists of the time, De Geer described American specimens that had been shipped to him across the Atlantic Ocean. De Geer's chief source for American insects was a Swedish pastor, Israel Acrelius, who resided in "New Sweden" (Delaware) from 1749 to 1756. It obviously took some time for De Geer to sort through the many specimens, as *Amblycorypha oblongifolia* sat in his cabinets, undescribed, for at least seventeen years. Of course, that's assuming someone else did not send it to him at a later date.

Charles, or Carolus, De Geer was born into a wealthy family in 1720 in the Swedish province of Ostergötland. His family had prospered in the iron mine and furnace business, and upon inheriting this, Carolus continued to maintain it as a profitable venture.

He grew up in the Dutch city of Utrecht, where he became interested in insects at an early age. It is said that interest was sparked by a gift of silkworms he received when he was eight years old. At sixteen he was studying aquatic spiders, a species known to spin their egg sacs in the water. By the time he was nineteen, he was elected into the Swedish Academy of Sciences. He was working on a paper on the Podura, or springtails, which he completed shortly after.

De Geer had great respect for Linnaeus and attended his lectures, but he did not always agree with his system of nomenclature, which, to be fair, was still a very new concept. In his well-known journal of his observations, the seven-volume *Mémoires pour servir a l'histoire des insectes* (Contributory Notes on the History of Insects), De Geer chose to model his work after René Reaumur. Reaumur was a French entomologist who had published a six-volume treatise on insects. It, too, was called *Mémoires pour servir à l'histoire des insectes.*[6] De Geer was a great admirer of the French scientist, although they had never met. The Swede even chose to write this journal in French. It was his expectation that readers would be more sympathetic to errors made by a man writing in a language he admittedly had to work a bit harder at. They were not. In fact, the first edition sold so few copics that De Geer burned much of the original manuscript.

That first edition, published in 1752, preceded Linnaeus's first edition of *Systema Naturae* by one year. I have perused a copy of it, and the species are named in the old manner of descriptions of various lengths. In following editions De Geer begrudgingly began to employ the Linnaean system, although he did not necessarily agree with it. He saw the writing on the wall, which told him the old system was going by the wayside. Many of the names of insects we see today, which were first described by De Geer, are courtesy of a publication following his death. In 1783 A. J. Retzius's publication *Caroli de Geer genera et species insectorum* reconfigured De Geer's nomenclature to fit snugly within the Linnaean system.[7]

It seems that De Geer was ever concerned that his peers in the world of science would find errors in his work. He qualified his potential for missteps by stating that his journal contained his own

Charles De Geer.

observations and was with "without excessive reasoning." In the beginning of the journal he explains that one needs to make a lot of mistakes before settling on the truth of a matter. It is difficult, however, to argue with observations. You see what you see. You describe what you saw. That was his approach. Much of what he saw was through a microscope, a relatively new tool for the natural historian of the day. This allowed him to describe things previously unseen. In addition to his writings, he produced over two hundred copper plates illustrating many of the nearly fifteen hundred species he described. De Geer was a keen observer of the form and life history of insects. His work was one of the earliest to share those observations in a scientific manner. He saw, like Linnaeus, that insects could be divided into categories based upon attributes of their wings. He added to that his own clas- sification that insects could also be grouped by the similarity of their mouthparts. This harkens back to the Aristotelian model. De Geer classified the Orthoptera in this way:

> Two wings covered by two wing-cases. Elytra are coriaceous [leathery] or semi-crustaceous, aliform [wing-shaped]. Pair of membranous wings. Mouth with teeth.[8]

That coriaceous-winged Oblong-winged Katydid chewing on the sassafras leaf would fit nicely in that category.

While I can appreciate the earthy forms of the crickets, I've always been more drawn to the katydids. Their song is in no way as rich as that of their cousins, but their form is a marvel to the eye. If leaf and stem were to rearrange their parts for the purpose of walking, a katydid would be the result. As I've mentioned, the species of katydid that sparked my interest for this order is *Scudderia furcata,* or the Fork-tailed Bush Katydid. The simple reason for this distinction is that this species was the one I first, and most often, encountered in my yard when I was becoming interested in the night-singing insects. It was also the first katydid I was able to reliably pin a call to, that being a short, lispy "tsip" given at varying intervals.

The "fork" in "Fork-tailed Bush Katydid" refers to the shape of the male's genitalia—more specifically, the double-lobed end segment of its abdomen. The shape of that segment, known as the *supra anal plate,* is how we separate the bush katydids from one another. *Furcata* is Latin for "shaped like a fork." "Bush" in "bush katydid" makes reference to their habitat, which tends to be shrubby and herbaceous areas along the edges of deciduous woodlands. The bush katydid family is composed of only eight North American species. They all share the genus *Scudderia.* Often a scientific name will reveal things about a creature that make it unique or interesting. I've long been in the habit of researching the meanings, or etymology, of the names of species I have wished to learn more about. At the time, the word *Scudderia* suggested to me some kind of crustacean. It also brought to mind "scupper," which is the drain on a ship. I wasn't even close. Because *Scudderia furcata* was one of my earliest introductions to katydids, Samuel H. Scudder, the source of the genus name, became my earliest introduction to the people who contributed to our knowledge of them.

Samuel Hubbard Scudder was born in Boston in 1837. He grew up among the woods and fields on a thirty-acre estate in nearby Roxbury. His parents were devout Puritans, each with American lineage going back to the early 1600s. Charles, his father, was a deacon in his church and made a living as a hardware merchant. While he was considered a cordial man, one can assume that he was not the inspiration for young Samuel's pursuit of natural history. In a story shared in a memoir written in 1924 by Alfred Mayor, Samuel recounts an experience he had at the age of ten. Upon finding and marveling at the

beauty of a stick covered in brightly colored fungi, he brought
it indoors to show his father. Charles looked at it, called it a
"dirty stick," and tossed it in the fire.

At the age of sixteen, having just entered Williams College,
Samuel was stirred at the sight of a butterfly collection at a
friend's house. What amazed him was the sheer number of
beautiful insects, all found within the immediate region of his
home. He made himself a butterfly net and collected butterflies until the arrival of autumn's killing frosts. By the time
he entered his junior year, he had decided upon a career of
entomology.

Through his many contributions to the journal *Psyche* I
get a true sense of Scudder's contributions to science. *Psyche*
is the scientific publication of the Cambridge Entomological
Club, of which he was a founding member. Members of the
club often met at Scudder's private laboratory, located at 156
Brattle Street in Cambridge. The greatest American entomologists of the day would sit in front of a cozy fireplace, engaged
in animated discussion well into the night. Scudder, by all
accounts, was considered an amiable man, fully unconscious
of his own charm. He had a leonine look to him, probably
suggested by his lantern jaw and long, bushy sideburns, which
nearly encircled his face. His expertise shone in three areas:
Lepidoptera (although he didn't seem to have much interest in
moths), Orthoptera, and fossil insects. He was wildly prolific,
which often led to mistakes. An amusing example of this is in
his description of the bristly leg of a fossil centipede, which he
called *Trichiulus*. It turned out to be part of a fern leaf. Later,
upon realizing his error, he suggested to friends that the name
be changed to "Tricky"ulus. Alfred Morse, whom we will
hear about later in this chapter, while having great admira-
tion for Scudder, wrote how his rapid work ethic might make
him prone to make more mistakes. In fact, he mentioned this

to Scudder, who replied, "Sure to! Sure to!" Part of the challenge faced by all biologists of the day was in dealing with and
making sense of the rapid intake of new information. These
were times of great curiosity and discovery. Naturally, there
was the desire to disseminate that information as it came to
bear. It is only over time, and with the subsequent acquisition
of more data, that corrections can be applied to errors made by
those pioneering scientists.

Scudder contributed 791 scientific papers to science. The
first was "List of Shells," which he wrote at the age of twenty-
one, in which he describes twenty-eight species of terrestrial mollusks. His last, written when he was sixty-three, was
on group characteristics of North American butterflies. He
endeavored to share the information he collected for a book
that he had given up hope on finishing. Although he lived
to be seventy-four years old, his work output suffered a rapid
decline following the death of his twenty-seven-year-old
son, Gardiner, to tuberculosis. The scientist had lost his wife,

Samuel H. Scudder.

Ethelinda, when Gardiner was only three. He never recovered in health or spirit following the loss of this son, and in the year his son died, Scudder developed symptoms of Parkinson's disease.

Scudder did manage to see to fruition what I would consider his greatest contribution to the study of Orthoptera, his *Guide to the Genera and Classification of North American Orthoptera,* published in 1897. Unlike some of his butterfly books, which were geared toward a more general naturalist audience, this was a highly technical guide covering the taxonomy of all the known genera of Orthoptera in North America. Aside from a poem by Keats, found after the title page, it contains not the faintest hint of pathos—not that it should, although one can usually find, at least in the introduction, some mention of the author's inspiration in creating the work. Instead, the introduction warns the reader that much work remains to be done on this group. What made this guide valuable is that it put in one place the culmination of the most up-to-date published, and yet-to-be published, findings of the time. Many of the names of the genera have been changed since that publication, but the grouping is similar to what we see today.

Many of the descriptions of his insects came from specimens he collected when he was younger. In addition to grasshoppers, crickets, and katydids, Scudder included earwigs, walkingsticks, roaches, and mantids in with the Orthoptera. Although some naturalists, especially outside of North America, still consider these to be Orthoptera, they are now generally thought to be in different categories.

I recently came across a short piece written by Scudder called "The Note of the Katydid." I was surprised that a man who seemed so interested in katydids thought so little of the call of one of its greatest heralds, the Common True Katydid. Of this insect, Scudder writes:

Since I began to study the character of the notes
produced by different species of Orthoptera, it has
been my fortune to hear that of the true Katydid but
once . . . The note, which sounds like xr, has a most
shocking lack of melody; the poets who have sung its
praises must have heard it at the distance that lends
enchantment; in close proximity the sound is exces-
sively rasping and grating . . . Since these creatures are
abundant wherever they occur, the noise produced by
them, on an evening especially favorable to their song
is most discordant.[9]

My thought on this, as someone who delights in the call of
the Common True Katydid, is that sometimes a sense—sound,
smell, taste—can be tickled in a favorable manner by the mere
fact of an early association with it. For me, the call of a true
katydid takes me back to summer evenings of my childhood,
a time when I could stay out late and play with my friends.
Staying out past "dark" was gift of summer. I don't think I ever
consciously noticed the calls accompanying that gift, but our
brains have a way of listening to things even when we're not
paying attention to them. Because Common True Katydids'
northern range falls more within southern New England,
perhaps a young Samuel missed out on hearing them, as he
suggests in the above piece.

In "Some American Crickets," Scudder describes a handful
of the calling crickets in his area.[10] His writing combines an
artful mixture of accessible prose and keen scientific observation. However, what strikes me the most is the prophetic
ending to the piece:

Such are a few of the commoner or more striking
forms among our American crickets. The study

of their songs has not gone very far, but enough is known to make it highly probably that we shall one day be able to distinguish each species that sings—and there are a very few that do not—by the character of the song, just as our birds are so recognized by the field ornithologist.[11]

That day has come.

Sadly, the last decade of Samuel Hubbard Scudder's life was consumed by Parkinson's disease. He was lovingly cared for by his sister-in-law, Miss Blatchford, who read to him every day. One of the books he requested of her was the original edition of *Dr. Johnson's Dictionary.* He wanted to hear every one of the 42,773 words and 114,000 quotations within. It seemed that the scientist's mind still craved that predilection for detail that had thrived in him for so many years. The man considered to be "the greatest Orthopterist America has produced" died on May 17, 1911.[12]

I've long associated Samuel Hubbard Scudder with the names of two other orthopterists. Both men had credited much of their knowledge of the Orthoptera to the work Scudder had done; both published books on Orthoptera in 1920; and both provided me with my earliest sources of information on this group. Their names were Willis Stanley Blatchley and Alfred Pitts Morse.

At the intersection of U.S. 19 and Lee Street in Dunedin, Florida, if you look to the right while traveling north on 19, you will see a marker erected by the Dunedin Historical Society in 1976. It reads:

> Dedicated to the memory of Dr. Willis Stanley Blatchley, 1859–1940. Distinguished naturalist from Indiana, who from his home across the highway

devoted twenty-seven winters, between 1913 and
1940, to the highly skillful exploration of the world
of nature around Dunedin and elsewhere in southern
Florida, writing some of his fascinating nature-study
books and parts of his classic volumes on orders of
insects.

Many of the entomologists I've discussed made contributions
to science beyond a single subject. Blatchley was no exception.
He wrote manuals on the Coleoptera (beetles), Heteroptera
(true bugs), Rhynocophora (weevils), and Orthoptera. He
also wrote many essays on a variety of topics ranging from
teaching school to religion. What originally brought him to
my attention was his book *Orthoptera of North-Eastern America*.
In my own book, *Guide to Night-Singing Insects of the Northeast*,
published in 2009, I mention his book in my foreword: "My
most treasured guide is *Orthoptera of North-Eastern America* by
Willis Stanley Blatchley, published in 1920. It's a big heavy
tome with a black cover, appropriate in appearance and weight
because it served as my bible for these insects for many years."
The book is 784 pages long, an inch and a half thick, and
covers 353 species of Orthoptera. It expands on his *Orthoptera
of Indiana,* which he published seventeen years earlier. Under
"Acknowledgements," he notes that he was greatly aided
by Samuel H. Scudder, whom he refers to as the "father of
American Orthopterology."

Unlike most of the work Scudder published on this topic,
however, Blatchley's tome contains a reason for becoming
engaged in these insects. He includes more of a "why" to seek
them out to Scudder's "how" to identify that which has been
sought and found. Interspersed within the intricate scientific
descriptions and keys are personal recollections of his and his
colleagues' hunts for crickets and katydids. These are fasci-

nating stories of scientists in the field searching for ways to describe their encounters. We read of them puzzling things out and coming to conclusions, and experience through them the joy of discovery. The crickets and katydids are no longer merely specimens in the lab. They are in context with what makes them what they are. In a review in the journal *Psyche,* the manual is called the first of its kind.

I have always believed that the best way to enjoy the denizens of the natural world is to experience them *within* that world. It's the difference between watching a bear in the zoo and seeing one lumbering through a forest. Blatchley found a balance of sharing the necessary scientific information needed to identify the insects and bringing them to life as best as words would allow.

One of the many species Blatchley described and named was *Nemobius* (now *Neonemobius*) *palustris,* the Sphagnum, or Marsh, Ground Cricket. In chapter 8 I wrote of my hunt for this species in a Connecticut bog. Blatchley's book is where I learned the technique for catching these diminutive insects. The author quotes Canadian entomologist Edmund Walker, who wrote, "They were present in numbers but were difficult to capture. By pressing the masses of sphagnum down under water it was often possible to bring the crickets to the surface." I tried this, and it worked beautifully!

In his autobiography, Willis Stanley Blatchley tells us he was born in October 1859 in North Madison, Connecticut (a town ten minutes from my own).[13] One year later his family moved to Indiana, where his parents bought a farm. They raised fruit and vegetables, and Willis worked on the farm during the summer until he was twenty years old. The farm gave him his first introduction to insects, as part of his job entailed collecting and eradicating potato beetles. For a few years he earned money by going door-to-door peddling clothing accessories (buttons, buckles, and such). He then took a six-week course to qualify as a schoolteacher and landed a teaching job in a country school for $1.50 a day. His summers were spent traveling on foot to sell maps and books, and taking orders for enlarging pictures. A year later, he took the job as principal in another school, earning him an income bump to $2.50 a day. Of insects, at that time he knew best the "bumble-bees, honey-bees, bald hornets and yellow jackets, all of which had, from time to time, forcibly impressed themselves" upon his perception.

At the age of twenty-four he entered Indiana University. He listed his assets of the time as three hundred dollars, a wife, and a baby; liabilities—zero. When he graduated four years later, his assets were listed as "same wife and two babies—still without liabilities." I find it hard these days to imagine graduating from college without liabilities, but to pay for his tuition he worked as a janitor, collected delinquent city taxes, gathered plants for the botany department, and continued to walk door-to-door selling maps and books.

Willis's original plan was to major in German, but he soon discovered his limitations in picking up foreign languages. He still had an interest in science and decided to meet with Dr. David Starr Jordan, who was head of the Department of Zoology. Dr. Jordan was busy working with some preserved specimens of fish when the young student came up to him. After asking Willis some questions, the teacher reached into one of the cans and pulled out a brilliantly colored fish. He said, "Well, you may take your choice between 'Ich bin, du bist, er ist,' and the study of such objects as this." Blatchley chose the fish and credits that moment with the beginning of his career as a naturalist. Three years later, while out on a "tramp" with other students, Willis found himself sharing a bed with his mentor. They had ventured too far and came upon a cabin. The owner reluctantly allowed them to spend the night. Until that point, Willis had viewed Jordan with a kind of reverence, but he said that something about sleeping beside someone in a strange cabin broke down such barriers. He confessed to his teacher that he did not want to study fish anymore and asked him for advice on a new direction. Jordan told him that entomology was one of the most neglected subjects in the country and there were thousands of species to be discovered. Willis took his advice.

From 1887 to 1894 he taught high school science, but his late afternoons and Saturdays were spent collecting insects. His summer peddler job was replaced by a variety of scientific pursuits, including the exploration of a volcano in Mexico, fish collector for the U.S. Fish Commission, and map work for the state of Arkansas. In 1894 he was elected to the position of Indiana State Geologist. He had concluded that this was a job he could do well, since the previous holder of that position was able to do it while frequently drunk. This station kept him outdoors a lot, something that was important to Blatchley. His interest in insects, specifically beetles, was growing, and during these years he began working on his manuscript on the Coleoptera of Indiana. He was often seen afield with his umbrella, which served as his most important collecting tool. The entomologist would beat branches with a stick, and the insects would fall into the umbrella for sorting. He also carried a notebook and took copious notes on his observations. In January 1908, on one of his hunting trips, he slipped on ice and broke both bones of his right lower leg just above the ankle. Blatchley wrote, "Crawled over ice and snow nearly one-fourth of a mile before getting aid; was on crutches for twelve weeks."

Blatchley began his book on the Orthoptera in the summer of 1917. Over the next three years he traveled between his home in Florida and Massachusetts. He spent two weeks in Cambridge studying Scudder's collections and a week in

Philadelphia, where the collections of Rehn and Hebard are housed. It took Blatchley less than three years to complete this work.

Following that, he authored a number of papers and essays. In his fourth and last manual, *Heteroptera, or True Bugs of Eastern North America, with Especial Reference to the Faunas of Indiana and Florida* (1926), he shared how his works were conceived of necessity and brought forth with great labor.[14] One of his justifications for this was that his "brain continued to demand employment of some nature."[15] This book did for the True Bugs what his Orthoptera manual did for that group.

In 1930 his failing eyesight and a chronic case of neuritis put an end to his work with insects. The last ten years of his life were spent wintering in Florida and summering in Indiana. He switched his studies back to the plants and general natural history of Florida, much of it in Dunedin, where his plaque now stands, and Indiana. He also took up stamp collecting. In 1937 Blatchley visited the home of his birth one last time. He returned to Indianapolis by "air-plane," which was the first time he flew, and marveled at the mere four and a half hours it took to get home.

Much of what we know about Blatchley comes from the man himself. He wrote about Willis Stanley Blatchley with the interest and introspection given his natural subjects. In addition to his biography, he published *Blatchleyana* and *Blatchleyana II,* which were year-to-year accounts of the highlights of his life from birth to the year before his death. He felt it important to document his accomplishments himself, so they would be recorded with accuracy. Blatchley died in his Indianapolis home in 1940, at the age of eighty.

I feel a connection to this man, born almost one hundred years to the day prior to me. I, too, traveled up and down the East Coast in search of Orthoptera to write a guide to their lives and identification. I spent much time chasing critters in Madison, Connecticut, the town of his birth. In fact, Michael DiGiorgio, the illustrator of our cricket and katydid guide, lives in that very same town and is likely to have found the species he contributed within scant miles, or feet, of Blatchley's own collecting spots. What I truly empathize with, however, is the ease with which Blatchley threw himself from one discipline to another. He had a restless mind that was allowed free rein in the fields and forests that called to him.

A contemporary, Albert Pitt Morse, credits Blatchley (and Scudder) with his contributions to his treatise, "Manual of the Orthoptera of New England."[16] Like Blatchley's book, this is a heavy tome, 594 pages, but bound in paper. It is similar in feel in that it shares the observations of many scientists and naturalists, who seemed to enjoy that gift for spinning a good nature tale. In fact, he begins this book with a quote from *On the Parts of Animals,* by Aristotle, one of the earliest naturalists:

> Wherefore we ought not childishly to neglect the study even of the most despised animals, for in all natural objects there lies something marvelous.

Albert Morse was born in 1863 in Sherborn, Massachusetts. He was a direct descendant of Samuel Morse, who in 1636 fled England to avoid religious persecution and difficult economic times. He settled in Dedham, Massachusetts. His son, Daniel, left Dedham to become one of the incorporators of the town of Sherborn in 1674. Albert lived on Charlescote Farm in a federal-style home built in 1759 by his grandfather. The home had replaced a fort his ancestor had built for protection during the Indian wars.

Albert graduated from the local schools at the age of sixteen. Because of his frail health and duties on the farm, it was the only formal schooling he would receive. He did, however, manage to find time to explore the natural world around him. At an early age he began to collect specimens and experiment with taxidermy. His outdoor pursuits were encouraged by local naturalists, among them William Edwards, professor of botany at Wellesley College and one of the founders of the Historical and Natural History Society of South Natick and Vicinity.

His association with Edwards likely played a role in Morse's accepting a position as an assistant in Wellesley College's Zoological Department at the age of twenty-three. He also credits M. A. Willcox, professor of zoology, with setting him on his path in the natural sciences. Willcox was the author of *The Pocket Guide to the Common Land Birds of New England* (1895). His interest in birds was shared with his student, who two years later published his own *Annotated List of Birds of Wellesley and Vicinity, Comprising the Land-Birds and Most of the Inland Water-Fowl of Eastern Massachusetts.*

Morse spent some time at Cornell University studying insects under John Henry Comstock, professor of entomology and invertebrate zoology. His training, Morse says, "led to many happy hours in the laboratories, fields, and ravines of Ithaca, and in after years in the wilderness trails of the far West and South."

Morse remained involved in Wellesley College for most of his life, working in many capacities. He married Annie McGill of Dover and in 1900 became the last Morse to reside in the old family homestead on Charlescote farm; they moved to Wellesley to be closer to his work.

In 1897 Samuel H. Scudder encouraged Morse to spend a summer collecting Orthoptera in the Pacific Northwest. Both men were active in the Cambridge Entomological Club and served in many roles, including president. For twelve weeks Morse explored the area between the Sierra Nevada Mountains and the Pacific Ocean. He traveled via the railroad, riding in a "second class" sleeper and occasionally staying in cheap hotels along the way. Sometimes he just slept under a blanket beneath the stars. When he needed to go farther afield, he would hire guides and pack mules to tote all his collecting gear. He returned with thousands of species, many new to science. It was Scudder, however, being the Orthoptera juggernaut of the day, who got to describe and name them.

Albert Morse did get to describe many new species, and he added to the records of known species that could be found in his native New England. His "Manual of the Orthoptera of New England" and Blatchley's *Orthoptera of North-Eastern America* together usher in a new genre of entomological literature—an invitation from the scientist to the amateur naturalist to join in the discovery, study, and joy of these creatures. These two men knew one another and generously contributed to each other's research. They were eager to disseminate the information that so excited them. Morse recognized the power of our aesthetic sensibilities and the influence the songs of insects can have on those who hear them. Of the tree crickets he writes:

> These insects are, in truth, veritable dryads, of fairy-like daintyness and evasiveness, often heard but seldom seen. (401)

He continues with a quote by the novelist Nathaniel Hawthorne, who describes the song of the Snowy Tree cricket as "audible stillness . . . If moonlight could be heard, it would sound like that" (402).

In 1934 Albert Morse's health began to deteriorate. His final contribution to science was a piece was about a goldfish-robbing kingfisher, written for the *Bulletin of the Essex County Ornithology Club.* He died in Wellesley in 1936.

The Cambridge Entomology Society (now the Cambridge Natural History Society) was a hotbed of world-renowned scientists in this field. A little over three hundred miles south, two men were toiling away at the Orthoptera from *their* neck of the woods. Their names are often linked: James Rehn and Morgan Hebard. Although preceded by them a couple of decades, they were contemporaries of Blatchley and Morse. Both men were fixtures of the Academy of Natural Sciences in Philadelphia and played a large role in making it the leading resource for Orthoptera research in the world.

One of the manuals that helped guide me in my own research of katydids is their "Studies in American Tettigoniidae," published in 1914 by the American Entomological Society. Their information comes from examination of thousands of specimens they collected through the country, which were, and are, housed at the Academy. They also perused the collections of a number of other entomologists, including the extensive holdings of A. P. Morse. They had the research of Scudder to work with, but found a number of errors they hoped to correct.

James Abram Garfield Rehn was born in Philadelphia in 1881. He had an interest in natural history at an early age, which is not an unusual trait for a scientist. Few in the natural science area fall into that field without having had such interest as youngsters. James was a member of the Aristotle Society, which was run by Charles Willison Johnson. The group was organized as an outlet for curious young minds craving more information about natural history. Johnson's main interest was in marine mollusks, but he also spent a lot of time in the field of entomology and was curator of the Wagner Free Institute in Philadelphia.

At the age of nineteen, Rehn became associated with the Academy of Natural History and joined the staff of the entomology department. He was also very active with the American Entomological Society. He was twenty-one years old when he first met the Cleveland-born sixteen-year-old boy with whom he would form a forty-year friendship and working relationship. Morgan Hebard had gone to the Academy looking for assistance in identifying some Lepidoptera he had collected. In a 1948 obituary for Hebard, Rehn says of that moment, "As a student then, some five years older, I gave him such assistance as was in my power and the acquaintance then formed ripened into a friendship and association which lasted through the years until his death, becoming to each of us a very vital part of our lives."[17]

In time, Hebard's interest in insects became focused on Orthoptera. He was born into a wealthy family, which allowed him, at the age of twenty-four, to devote his life fully to the study of this group. His family had made money in the lumber industry. At one time they owned five hundred acres that encompassed nearly the entire Okefenokee Swamp in southern Georgia. The land was eventually transferred to the United States as a National Wildlife Preserve. This transaction turned out to be a godsend for me, as it was one of my most productive areas in my own hunt for crickets and katydids.

Early in their relationship, the two young men embarked on a goal to create a comprehensive series of orthopteran field studies covering all the major areas of the United States. This would be presented in a monograph of the Orthoptera of North America. Their main thrust was to acquire information on this order in places undisturbed by human development. Cultivated land was bypassed, as they were not interested in the orthopteran by-products of land cleared of native vegetation. This led them climbing hand-over-hand up North American mountaintops and trudging through the most arid deserts. At the onset, they, like Morse, followed railroad lines, stopping and exploring roughly every fifty miles or when they passed through a distinctively different habitat. At times they rode in horse-drawn wagons or walked alongside pack mules to get into the backcountry. They camped out in tents or under a single blanket. Eventually the "era of luxury" entered, allowing them the opportunity to travel by "motor truck." The truck, packed with all manner of equipment, would resemble that of the Joad family as they left Oklahoma in the movie *The Grapes of Wrath*.

Much of this was funded by Hebard, who was a curator at the Academy but accepted no salary. He was also the financial backer of many of their publications that came out over the years. Data was released in the form of scientific papers and manuals throughout their forty-three-year association. Though they were often collaborators, they would also publish on their own. I am fortunate to have in my possession *The Dermaptera and Orthoptera of Illinois* (1934), authored by Morgan Hebard. It is an exceedingly thorough treatment of this group, covering everything from habitat and life histories to eradicating those deemed pests. One can get a sense of the motivation driving Hebard in his opening paragraph, in which he writes:

> The Orthoptera first attract one's attention because of their universal presence and relatively large size . . . Some are very drab and sullen creatures, but others flash brilliantly colored wings or crackle stridently through the air. A closer acquaintance however discloses that these bold forms which first catch the eye are no more interesting than a host of their allies . . . who hide away in their surrounding and must be sought with care and patience. It is among these that we find not only the fanciful and bizarre form, but also the finest assemblage of musicians produced by the insects.[18]

Hebard was described by his friend as an indefatigable worker, endowed with superb hearing, which was of exceptional value in locating nocturnal stridulating forms. He was inexhaustible in the field, and in the lab he possessed such a high degree of concentration that he was almost oblivious to what went on about him. He hated meetings, deeming them a waste of time, but, like Rehn, was active in the American Entomological Society.

In the 1930s Hebard began to suffer from the onslaught of arthritis. As the disease progressed, he spent long periods of time seeking relief in hospitals in Boston, New York, and Philadelphia. In 1946, when he finally reached a point where he could be somewhat mobile again, he died of a heart attack. One year prior, he had donated his entire collection, comprising 250,000 specimens of North American Orthoptera, to the Academy. Added to that were the thousands of species from the far reaches of the world.

A lot of what we know of Hebard's life comes from the obituary written by Rehn for the *Entomological News* in 1948. He laments the loss of a friend who formed a very vital and inseparable part of his own days. Rehn continued to travel the world in search of insects to bring back to the museum. In addition to his work sorting out the morphological characteristics that separated species and genera, he studied and recorded every aspect of their life history.

James Rehn stayed with the Academy until the death of his wife in 1964. I've had the opportunity to look at his personal notes, which fill twenty volumes housed in the entomology department. He has left behind reams of paper filled with his neat handwriting, noting details of where and when he found what. Over 350 papers bear his name either as the sole contributor or as joint author.

In 1961 the monograph treating all of the Orthoptera of

North America, conceived by the two Academy colleagues nearly forty years earlier, saw the light of day. Only volume 1 was produced. Rehn had teamed up with Harold J. Grant Jr., and their 257-page work covered the Tetrigidae, Eumastacidae, Tanaoceridae, and Romaleinae.

Rehn, whose decline in health was undoubtedly hastened by the death of his wife, died in 1965. Grant continued to work on the next volume, but died two years later. Daniel Otte, associate curator of entomology at the Academy of Natural Sciences of Philadelphia, credits, among others, Rehn and Hebard for their collecting and research done for the Academy. He calls volume 1 of his book *The North American Grasshoppers* a new beginning, as opposed to a continuation of his predecessors' monograph on the Orthoptera of North America.[19]

There have been, in fact, many new beginnings of late—in book and Internet format—covering the Orthoptera. Perhaps every turn of the century brings a new wave of interest in that area.

ten

Assembling Your Cricket Radio

Seven-year-old Caleb Brand needed a little help falling asleep. His father, Andy, who happened to be a naturalist, introduced a unique solution: singing insects. He and his son would go out in the yard with flashlight and net, and shop for potential roommates. For them, the hunt was as much fun as the procurement.

The insects were set up in comfy quarters of their own and given all that was necessary to provide them with a reason to sing Caleb his lullabies. His summer nights were tuned in to "Cricket Radio."

A number of different species were booked for the lineup. Eastern Swordbearers *(Neoconcephalus ensiger)* and Long-beaked Coneheads *(N. exiliscanorus)* would provide their gentle "swishing" calls, sounding like straw brooms sweeping steady rhythms over a cement sidewalk. Field crickets *(Gryllus* sp.) and Snowy Tree Crickets *(Oecanthus fultoni)* became chirping metronomes, their songs marking the passing minutes. Common True Katydids *(Pterophylla camellifolia)* croaked their names into the wee hours. At one point they'd captured a Greater Angle-wing *(Microcentrum rhombifolium).* This species makes a sound akin to two pebbles being tapped together; to this it will often add a series of lisping "tzips." Andy and Caleb brought it to a party

Greater Angle-wing *(Microcentrum rhombifolim).*

we were attending at a friend's house. Andy was pretty sure he knew what it was, but wanted to confirm the identification. My jaw dropped when I saw it. At the time, I was working on my night-singing insect guide, and before me was a species I'd been trying to catch for years. I needed to record its call and take some photos for the artist's reference. The Greater Angle-wing is an arboreal species one rarely gets to see. They spend their entire lives in the tops of deciduous trees. And here it was, in all its leafy splendor. It perched calmly on a little branch in a translucent plastic container. Andy had modified the container by cutting a hole in the lid and covering it with a screen.

I asked him how attached his son was to this insect. Andy said, "Pretty much." Rats. I was kind of hoping he'd say something like, "Eh, he could take it or leave it." I asked him if there was any way I could borrow it for a couple of days. Andy, as he should have, informed me that it was up to his son. Caleb was hesitant, but agreed, as long as we could find something to take its place. Fortunately, the venue of this party provided good potential for finding insects. We all headed out into the fields in search of a replacement singer for this angle-wing.

It was Caleb who came upon a candidate. I looked at the little, green, half-inch bug in disbelief. Here was yet another insect I had wanted to see—this one since I was Caleb's age! It is featured on page 33 in *A Golden Guide: Insects* by Herbert Zim and Clarence Cottam. I still have the book. It was my most valuable possession as a child, and still holds a special place in my bookshelf.

This bug obviously held a special place in the heart of the illustrator, James Gordan Irving, since he gave it its own page. His Buffalo Treehopper *(Stictocephala bisonia)* stares at you, looking like a bull ready to charge. The insect is rendered in three-quarter view, and blown up about fifty times its actual size. It has an imposing look, with its arching humpbacked wings, but at the same time holds a gentle quality. It's a leaf mimic, and how tough can a leaf look?

Caleb seemed to be taken with it, too, as was Andy. The problem was, leafhoppers don't sing. Caleb didn't care. He said I could have his angle-wing—we were even. I really felt guilty about replacing his singing insect with one that is mute. However, I could see he was going home with something that spoke to him in a different way. I could see that, because it did the same for me.

I fully intended to return the katydid when I was through with it, anyway. But it was not to be. It had escaped in my yard as I was photographing it.

That was the only male Greater Angle-wing I've ever seen. I have come across a few females. They call too, but to a lesser extent, and they need a male's call to bring it out of them.

The ancient Chinese are known to have appreciated the songs of insects in the way Caleb had. One of the earliest books on the subject of cricket keeping was written in the first part of the thirteenth century. Kia Se-tao, a minister of state, wrote *Tsu Chi King* (Book of Crickets), in which he gives descriptions of the calls of different species, as well as advice on their treatment and care. This book, however, does not represent the beginning of this practice, which goes back two thousand years. Bethod Laufer, in his book *Insect-Musicians and Cricket Champions of China,* shares a quote from the Chinese book *T'ien pao i shi* (Affairs of the Period T'ien-pao, AD 742–756):

> Whenever the autumnal season arrives, the ladies of the palace catch crickets in small golden cages. These with the cricket enclosed in them they place near their pillows, and during the night hearken to the voices of the insects. This custom was imitated by all people. (18)

Other ancient cultures kept crickets and bush crickets (katydids) for their song as well. There are references, also over two thousand years old, to this custom taking place in Greece. Japan, Germany, India, and many other countries have records of this practice in their long histories. Many continue to do this today. However, here in the United States it hasn't really caught on.

The easiest way to invite insect song into one's home is to

Brass cricket house.

simply open the windows. The concerts taking place outside our homes are surely there for the listening, free of charge. There is, though, a certain pleasure to be gained in choosing one, or several, of those songsters for a private performance.

I do like the idea of adding to the aural tapestry of one's home. A sound can provide a certain accent to a room, just as effectively as color or fragrance. Some people hang wind chimes out on their front porch for just that reason. Some enjoy the hourly call of a little wooden cuckoo residing in a clock on their wall, or the deep ticking of a grandfather clock. My wife says the ringing of a buoy bell brings her back to her childhood nights of sleeping at her family's summer cottage on the beach. That is often the case with the enjoyment of certain sounds. They bring us back to a happy time in our life. And isn't this the essence of music? Doesn't the music we choose to listen to trigger emotion?

Some time ago I was reawakened to the night sounds, to the point where I wanted to bring into my home some of the sounds I was hearing outdoors. The way I see it, it's no different from going to a pet shop and bringing home a songbird. Okay, it's a *little* different. A pet bird is a years-long commitment. A cricket doesn't live much longer than a few months, and even then I don't hold on to them for much more than a week or two.

I started with a Spring Field Cricket *(Gryllus veletis)*. They were plentiful and accessible, and had a familiar call. First I procured a single male from beneath a log in my yard. I set up a mini terrarium, my goal being to provide the guest with accommodations so close to what it was used to outdoors that it wouldn't know it was indoors. I wanted a happy cricket. And it seemed happy, adding its rich chirp to my studio day and night. After a couple days, though, I began to feel bad for the cricket,

Male Spring Field Cricket *(Gryllus veletis)*.

singing away, oblivious to the fact it had zero chance of calling in a female. While the song was pleasant, the state of futility took away some of the allure.

So, being a good host, I added a female to the cage. This brought a whole new element to the cricket's lone concert. Singing insects have a variety of calls, as do many creatures that produce sound. There are calls to attract from a distance, calls to initiate mating, and calls to repel competition and other threats. Within seconds after I'd added the female to the male's cage, his tune changed dramatically. It went from a steady, reedy chirp to a series of rapid, buzzing "chinks" as he sidled up to her, attempting to induce her to acquiesce to his charms. This provided for some interesting insight into cricket behavior.

I have been inviting the Orthoptera into my home for many years now. While I don't keep them in my bedroom, a summer never goes by without my adding the songs of crickets and katydids to my studio. This turned out to be a good way to place the song with the singer, too. Many of my Ensifera guests started out as unknowns, hearing their songs for the first time in my home. Watching them stridulate helped imprint in my memory the sound with its source. Now when I hear insect calls outdoors, I instantly visualize the cricket or katydid that is creating it.

Over the years I have learned a few things about keeping these insects in captivity. The most important lesson is that a happy bug is a singing bug.

Many different kinds of cages were used throughout history to house the singing Ensifera. Most seem designed more to please the listener than said insects. I prefer to create a minia-ture environment in which my captive can go about its time spent with me with as little change to its natural surround-ings as possible. That is the deal I've made with them, and with myself. I already feel a twinge of guilt when I remove a creature from its outdoor home. They belong where they were born and gain little benefit in being removed from there. Yes, there are more dangers out in the real world, but that's supposed to be a part of their life—and death.

I believe you can have something that is both an attractive part of your indoor landscape and a healthy environment for its inhabitant. This is not unlike having a fish tank. Many enjoy the visual aesthetic appeal they offer. The aquarium becomes part of the furniture and provides living entertainment. The same blend of art and natural science can be created with your singing insects. I would say it would go even beyond that, because in addition to the visual appeal, it offers a gift for the ears. Although I suppose many enjoy the sound of a bubbling fish tank.

To further assuage my guilt in kidnapping these singers from their homes, I make a point to release them back where they were found. I rarely keep an insect more than a couple of weeks. Most have a life expectancy spanning just a few months, so even a week removed from their home is a good chunk of their life. Exceptions are made for species I've brought from far locales, where returning them would be too difficult. Foreign species should not be released outside of their range. For example, when I brought back a male and female Japanese Burrowing Cricket *(Velarifictorus micado)* from the southern states, I had to let them live out their natural lives in captivity. Chances are good that our winters have prevented this introduced species from settling into the Northeast, but one never knows if some hardier stock could get them started up here. However, I did release the male after the female died. There was no chance of him furthering his progeny on his own in a Connecticut yard. He called for about two weeks from my meadow. The sound was oddly out of place.

Do these insects really know they are in captivity? Probably not, especially if things are done correctly. But *I* know.

Different species and families have different requirements for care. Again, my goal with all of them is to replicate their natural habitat as closely as possible. To start with, I use plastic aquariums or terrariums of various sizes. These are available at most pet stores. The larger insects will obviously warrant a larger container, as will the inclusion of more individuals within the container. These containers have transparent walls and a clear bottom. The lid is perforated with slots for air exchange; in fact, there is about as much area in openings as there is solid lid. Many have a clear, plastic lid *within* the lid, which makes it easier to drop in the insects and add or remove food without inviting escape.

I also use a large variety of jars to house these insects. Since most jars are more tall than wide, they are more often used

Rectangular container.

for insects that pass their time on branches, stems, and leaves, as opposed to on the ground. The lids need to have a lot of holes added for air and moisture exchange. One of the key concerns with keeping insects in a glass jar, or any container, is the buildup of mold. If there is no way for moisture to escape, then mold quickly grows on the plants, food, and ultimately the insects themselves. This occurs in well-ventilated containers as well. If mold is allowed to build up, the bugs die.

Whatever the container, it needs to lend itself to cleaning. Droppings have to be removed regularly, and uneaten or rotting food needs to be replaced. This means that you will need to have a second container handy—used as a waiting room for your guest while his place is being cleaned.

You may wish to consider having two cages per insect(s). This will allow you to transfer them into one while you clean the other. I've found it easier to have the second cage completely set up so I can make a quick transfer, and not worry about a second transfer to the newly cleaned cage. Escapes happen when moving the insects from one container to the other, so the less you have to do that, the better.

The bottom of the container is sometimes lined with a paper towel. This allows for an easy cleanup, because it just needs to be removed and thrown away. It can also hold a little moisture, which the insects require. Sometimes I just forgo the paper towel and leave the floor of the enclosure bare. In the past I would cover the bottom of cricket enclosures with soil. While I haven't had any problems with this, there is the potential of introducing predators, so I no longer do that. You could buy "clean" potting soil. This would eliminate the predator problem and give the setup a more natural look.

Because I am not interested in rearing these insects, I do not need to provide egg-laying substrate. Many laboratories rear insects for a variety of purposes and will provide sterilized sand or peat for ovipositing. My purpose for keeping the insects is purely for their song. There are always plenty of singing adults out there to find, so I've never found a need to raise them from eggs.

Water can be provided in a number of ways. It can be poured into a shallow cap, though one must be sure to keep the water below a level that could lead to drowning. Bird waterers, which consist of an inverted container and a spout, have also been used for crickets. I prefer to supply it in droplet, or dew, form, misting the inside of the aquarium or jar with water from a spray bottle. You can also give the leaves, and the bugs, a little shot. Once in the morning seems to be enough, but if it is warm out, there may be a need for a second spritz later in the day.

The setup should be in a place that allows the insect to experience the natural phases of light and temperature throughout the day, although the temperature may be more difficult to replicate. It gets pretty hot in my studio during the summer, and I sometimes run an air conditioner. This undoubtedly makes their environs cooler than the outdoors. However, many of the insects I will be talking about do seek out the shadows and cooler microclimes in those hot days. For this reason, you do not want to leave your cage where the sun's heat and rays will be inescapable. Ideally, it should be near a window, but not on the windowsill facing the sun, where the insects will bake!

I have divided the rest of this chapter into groups and individuals, and how to care for them. Included are examples of some of my favorite singers and those I have had personal experience with.

Field Crickets

Field crickets represent crickets on the larger end of the size scale. A couple dozen species and close relatives are found in

North America. These are hardy insects that do well in captivity. Most have a rich, loud call that is repeated often. Many will call throughout the day as well as night.

Care must be taken when putting two males together. As I wrote in Chapter 6, many species of male field crickets are prone to fighting. If you want to hear the chorus of several males, you can put them in separate, nearby cages. You can also place them around the room for surround-sound effect! That's not to say that I haven't housed male field crickets together. If they have enough room to get away from one another, there is less of a problem. The upside is that the closer proximity inspires them to call more frequently, and fervently.

Adding a female, or two females, should induce the males to provide their courtship calls. These can be shorter, and more intimate, than their main call.

Songs and Range

Spring and Fall Field Crickets (*Gryllus pennsylvanicus* and *veletis*): Rich, loud "chirp . . . chirp . . . chirp." It is a familiar call to most people. These are a very common species heard in our yards. FALL FIELD CRICKET RANGE: throughout the United States, except in the far Southeast. SPRING FIELD CRICKET RANGE: the Northeast, north of the Carolinas.

Sand Field Cricket *(Gryllus firmus):* Rich, loud "churp . . . churp . . . churp." The call is somewhat deeper in pitch than that of the Spring and Fall Field Crickets. A sand-loving species. RANGE: up and down the East Coast.

Eastern Striped Cricket *(Miogryllus saussurei):* A slow, questioning "zeeet?" given in several-second intervals. RANGE: these smaller members of the field crickets make it into the lower northeastern states and west to Nebraska.

Southeastern Field Cricket *(Gryllus rubens):* A rich, sustained, stuttery trill, "zeeeeeeet . . ." RANGE: southeastern states.

Japanese Burrowing Cricket *(Velarifictorus micado):* A series of burry chirps, "cheer-cheer-cheer . . ." An introduced species. RANGE: central to southern states, as far north as New Jersey.

House Cricket *(Acheta domesticus):* A high, rich, and burry "cheerp . . . cheerp . . . cheerp . . ." If they're not in your house now (and they probably aren't), you can buy them at pet stores. You may have to pick them out, because some pet store workers don't know the males from the females. There also tend to be a lot more females in the group. RANGE: eastern half of the United States.

Home

Because these are ground dwellers, I like to provide them with room to crawl on the bottom of the container. The rectangular aquariums suit them nicely. Field crickets will take advantage of the opportunity to go under things. A piece of bark with a bit of a curve to it works very well. If you use a few of them, you'll be able to create several nooks for them to hole up in. Give the bark a good rinse under hot water before adding it to the container. You can also use cardboard egg separators to provide hiding places.

Sometimes, to add to the cricket *feng shui,* I include some small stones and moss. Care must be taken, however, that the stones can't tip and crush the cricket. I learned that one the hard way (actually, the cricket did).

Food

Field crickets are omnivores. Lettuce and spinach leaves always go over very well and can supply the vegetable component of their diet. Avoid using the outer leaves, since that may have traces of pesticides. Add to this crushed dry dog or cat kibble, or dried fish flakes. They will eat most fruit, but I prefer using apples because they can be sliced thinly and placed, peel side down, on the bottom of the cage. Cucumbers, melons, squashes, and carrots are other favorites. Small, dead insects will be relished! Crickets also feed on a variety of grains, so oatmeal and bread can be added to the mix.

Because crickets are raised as pet food (for reptiles and amphibians), there are a variety of commercial products available for feeding them. You can buy jars of moisture and moisture/nutrient products to add to their sustenance. They come in a gelatinous cube form and are available in pet shops. The crickets take to them very well.

Water can be provided in a shallow dish. An occasional spritz from a spray bottle works well, though, and can provide added moisture to the overall environment.

Ground Crickets

Ground crickets are similar in appearance to the field crickets, but considerably smaller. Most have bristles on the pronotum, while the field crickets are smooth. Their calls are higher in pitch and quieter than those of the field crickets. I find the sounds they make agreeable to the ear. Because they are quieter, the songs seem to fall more into the background than those of their larger cousins.

The smaller size of these crickets allows for the inclusion of more males in the cage. With more room to get away from one another, there is apt to be less aggression. I've never found the males in this group to be as prone to fighting, anyway. Their warning chirps seem to be well heeded. The blended calls of several males can be even more enjoyable than the call of a lone cricket.

A couple dozen species are found throughout the United States.

Songs and Range

Striped Ground Cricket *(Allonemobius fasciatus)* and **Southern Ground Cricket** *(Allonemobius socius)*—A high, burry "chit . . . chit . . . chit . . ." The two species are closely related and identical in appearance and call. They are found in sandy habitats. STRIPED GROUND CRICKET RANGE: northern half of the United States. SOUTHERN GROUND CRICKET RANGE: south of Pennsylvania and west to the lower central states.

Allard's Ground Cricket *(Allonemobius allardi)*—The call consists of a high, sustained trill broken by brief pauses. RANGE: throughout much of the United States, except the far South and West.

Tinkling Ground Cricket *(Allonemobius tinnulus)*—A steady train of "tink-tink-tink . . ." It is my favorite call of the ground crickets—like a little glass bell. RANGE: most of the eastern half of the country.

Spotted Ground Cricket *(Allonemobius maculatus)*—A very high, fast, pulsing, and burry trill, "ti-ti-ti-ti . . ." RANGE: most of the eastern United States.

Cuban Ground Cricket *(Neonemobius cubensis)*—A high-pitched trill. RANGE: eastern states, south from New Jersey.

Sphagnum Ground Cricket *(Neonemobius palustris)*—The Sphagnum Ground Crickets have an soft, ethereal trill, making them a favorite on the cricket radio hit parade. Catching them is half the fun, as described in Chapter 8. RANGE: open sphagnum bogs throughout the eastern half of the country.

Carolina Ground Cricket *(Eunemobius carolinus)*—A continuous "riiiiiiiiiiiiiiiii," the "i" in "ri" forming a soft vowel sound. A very common species. RANGE: throughout most of the country but absent from the far Northwest.

Home

Ground crickets require the same housing as field crickets, but because they are considerably smaller they will not need as much room. Care must be taken that the openings in the top of the aquarium are smaller than the crickets. Some of the commercial containers could present a problem in this regard. This can be easily remedied by stretching a paper towel across the top before snapping on the lid. For a more permanent solution, cloth screening could be glued to the lid's inner surface.

Don't let their diminutive size fool you—they can jump! More than one has escaped by jumping onto the back of my hand when I was replacing food, using it as a launching pad to freedom.

Food

The feeding requirements for this group would be the same as those of the field crickets, with one exception. Sphagnum Ground Crickets feed on the sphagnum moss in which they live. Supplement this with crumbled dry dog or cat kibble, sprinkled right into the moss.

Trigs

Trigs, being climbers, can be more of a challenge to keep as pets, but their song makes the extra effort worth it. They are similar in appearance to ground crickets, but show more variety in color.

Some species, like the Colombian Trig of the Southeast, can be difficult to capture because they are up in the leaves of deciduous trees. This is unfortunate, since they have such a sweet call. Say's Trigs can be found by sweeping a net in grasses and low vegetation along the edges of freshwater habitats. The holy grail of this subfamily, the Handsome Trig, is found in similar habitats, just a few feet higher in the plants.

Songs and Range

Columbian Trig *(Cyrtoxipha columbiana)*—A series of high, steady, soft, tinkling trills. Males sing in unison with other males, creating a ringing chorus. RANGE: southeastern United States.

Say's Trig *(Anaxipha exigua)*—A high, breathy trill. RANGE: eastern half of the United States.

Handsome Trig *(Phyllopalpus pulchellus)*—A scritchy, high, stuttery trill. With a deep, cherry-red head and thorax, cream-colored legs, and a blue-black beetle-shaped back, this is one to seek out for, as the name suggests, its handsomeness! RANGE: eastern half of the country, but absent from the more northern states.

Home

These are small insects, so you won't need a large cage. However, they will climb up the sides and will crawl upside-down beneath the lid. This means you will have to make sure that they will not be able to slip through the ventilation holes. The holes in most commercial aquariums would be too large for these insects—at least for the Say's and Columbian Trigs. This can be remedied by gluing a fabric screen to the inner lid. In the past, when I was too lazy to go the screen route, I would just stretch a paper towel across the top of the container, and place the lid on top of it.

A jar would work very well, but instead of a metal or plastic lid, cover the mouth with fabric screening and hold this in place with a rubber band.

If you add some twigs and leaves to the habitat, they will have something to crawl on besides the underside of the lid, which is where they always seem to head.

Food

Trigs eat a wide range of food, including leaves, flowers, fruit, insect eggs, and very small insects. It's always good to pick a few leaves and flowers from the plant upon which they were found. Take some of the plant parts from the flora in their immediate vicinity as well. As with most of the insects I've kept, these crickets welcome lettuce.

Aphids, scales, and other tiny plant-crawling insects can be added. Whatever the insect, though, it should be one that climbs, as opposed to species that will hide on the bottom— under something. If they are in the vicinity of the crickets in their cage, there is a better chance they will be captured.

Spritzing with a water bottle is a bit more risky with the trigs, because they are usually very quick and will take advantage of the opportunity for escape. I'd swear there's a wiliness to them. However, since they are usually near the top of the enclosure anyway, you can spray through the screening without removing the lid (should you be using one). This will give them the moisture they need.

Tree Crickets

Tree crickets have a long history of being kept for their song. They are delicate in stature and found off the ground among leaves, flowers, and twigs. The songs are given in the form a trill, either continuous or broken up into shorter pulses. The volume varies, but many call surprisingly loud, considering their diminutive size.

These insects are very easy to handle, too. When cleaning out a cage, one needs just to transfer into a holding container the leaf or twig the cricket is on. The crickets don't recognize this action as a chance to escape because their platform is still beneath their feet. This has allowed me to get a little cocky, and I'll sometimes just rest the leaf bearing the bug on the table while I'm cleaning the enclosure.

There are about twenty species in the United States. I have had the pleasure of hosting nine of them.

Songs and Range

Narrow-winged Tree Cricket *(Oecanthus niveus)*—A soft, steady, pulsing trill, similar to the ringing of a phone. It is a comfortable sound I enjoy hearing inside and outside. RANGE: eastern half of the country.

Davis's Tree Cricket *(Oecanthus exclamationis)*—The call is similar to the Narrow-winged's, but the trills are less consistent in length. RANGE: eastern half of the country.

Broad-winged Tree Cricket *(Oecanthus latipennis)*—A very rich, loud, and continuous trill. It is an attractive insect with raspberry red on top of the head. RANGE: eastern half of the United States.

Four-spotted Tree Cricket *(Oecanthus quadripunctatus)*—A rich, continuous trill. RANGE: widespread throughout the United States.

Black-horned Tree Cricket *(Oecanthus nigricornis)*—A rich, continuous trill. I am unable to distinguish this insect's call from that of the Four-spotted. RANGE: northern states.

Fast-calling Tree Cricket *(Oecanthus celerinictus)*—A rich, continuous trill, sometimes interrupted by brief pauses. The name makes reference to the rapidity of the pulsing that produces the trill. RANGE: southeastern quarter of the United States.

Snowy Tree Cricket *(Oecanthus fultoni)*—Rich, evenly spaced chirps. The call is a familiar aural backdrop to a summer night in the country. RANGE: throughout the United States except for the Southeast.

Pine Tree Cricket *(Oecanthus pini)*—A rich, high, continuous trill. This call would be emanating from conifers. It is a very handsome insect with a pleasing call. RANGE: much of the eastern half of the country.

Two-spotted Tree Cricket *(Neoxabea bipunctata)*—A dry, buzzy trill, broken with short pauses. Sometimes the trill is continuous. The call seems to have a softer pitch when it is mixed with other chorusing males. Found high in the trees. RANGE: eastern United States.

Home

Tree crickets are climbers. They are rarely found on the ground, and on the rare occasion when they are, they will in short time make their way back up into the branches. You could use a rectangular aquarium to house this species. Branches with the leaves attached can be placed along its length. Jars, which provide more vertical area, work well, too. The only drawback of using a jar is that the glass can make the call sound a little "ringy." This is more likely to happen when the cricket is calling from the lower area. You can perforate the lid of the jar to allow for gas exchange, but it would be better to cover the top with a fabric screen held in place with a rubber band. This better allows the call to escape, and can slow down the mold growth.

I have been able to find cylindrical terrariums in pet stores. These are made from clear, hard plastic, and the lid is highly perforated. They are perfect for keeping tree crickets.

The best combination for your terrariums is one male and two females. While most crickets do not need the female present to call, they do call more frequently, and with a bit more fervor, when they are not alone. Many tree cricket species join in chorus, so having a number of males in the room could make for a nice effect.

Cylindrical container.

Food

The mainstay of a tree cricket's diet is flowers and leaves. They will feed on the flowers of a large variety of herbaceous plants, including, but not limited to, goldenrod, buddleia, daisy, black-eyed susan, purple coneflower, and the flowering parts of fruit trees and shrubs. They also eat leaves such as cherry, sassafras, birch, apple, maple, plum, and grape. Whatever leaf you find one on can be considered a good candidate for a meal. Small, round holes chewed through the inner area of a nearby leaf can be an indication of tree cricket feeding. As with nearly all of the Orthoptera, they will readily accept lettuce as food.

Pine Tree Crickets are the exception. They feed on the needles of conifers.

Choose the young, soft leaves and snip off the branch, leaving them attached. This will keep the leaves fresher for a longer period. Rinse the leaves in cold water and pat them down with a paper towel to remove any tiny predators that may have hitched a ride.

Do not rest the stem in an open water container—the crickets can fall in and drown. You could rig up a paper towel or plastic cover to prevent this. I use florists' stem tubes to provide water for the plant cutting. These are the little plastic vials shaped like test tubes that hold single-stem flower cuttings. There is a plastic cap on the top, with a hole to insert the stem.

Stem holder.

The hole is small enough to tightly encircle the stem, preventing leakage. The last option is to simply replace the leaves when they begin to dry out.

Tree crickets also feed on other insects, their favorites being aphids. I've found that the easiest way to add aphids to their cage is to clip the branch the aphids are feeding on and lean it against one of tree cricket roosts. The crickets will actively hunt them down and eat them! They also eat scale insects, leafhoppers, gnats, and many other small, soft-bodied insects.

Just about any fruit can also be added to the mix. I've found apples to be the easiest to work with. Allowing the slice to rest on the peel keeps it from sticking to the bottom of the cage when it's time to clean it out. The apple also tends to hold up a little better after freezing. You don't have to waste an entire apple to provide a single sliver for one of the insects. It can be sealed in plastic and stored in the refrigerator or freezer.

Because tree crickets are more likely to hunt the branches and leaves for food, dog or cat kibble sprinkled on the bottom of the cage may go unnoticed. They do, however, seem to find the fruit. I sprinkle the crushed pet food or fish flakes onto that fruit so they won't be able to miss it. The added protein helps meet their nutritional needs.

For water, just give the leaves and walls of the container an occasional misting with a spray bottle.

Mole Crickets

I love the call of the Northern Mole Cricket! Its burry "dirt-dirt-dirt" strikes a particular pitch that hits my ear just right. The calls of the other mole crickets vary in tempo, but share that burry quality. The songs of the Northern and Prairie Mole Crickets, to my ear, are the ones with more listening appeal.

Habitat-wise, they are the complete opposites of the preceding tree crickets. They dwell in burrows beneath the ground's surface, but at night and in early afternoon they come to the surface to feed on plants.

The males will call without females present, because in the wild they wouldn't know the females were in the area anyway. There are seven species of mole crickets in North America, most of them accidentally introduced.

Songs and Range

Northern Mole Cricket *(Neocurtilla hexadactyla)*—A burry, croaking, steady series of "dirt-dirt-dirt . . ." RANGE: eastern and central United States.

Prairie Mole Cricket *(Gryllotalpa major)*—A burry, steady series of "dirt-dirt-dirt . . .", higher in pitch than the Northern Mole Cricket. RANGE: central United States.

Southern Mole Cricket *(Scapteriscus borellii)*—A continuous burry trill, similar to that of the Broad-winged Tree Cricket. RANGE: southeastern states, west to Texas.

Tawny Mole Cricket *(Scapteriscus vicinus)*—A continuous burry trill, higher in pitch than that of the Southern Mole Cricket. This one sounds like a Black-horned Tree Cricket. RANGE: far southeastern states, extending west to Texas.

Home

Mole crickets need a few inches dirt in which to burrow. The easiest way to obtain this is to scoop up the dirt your cricket was originally found in. I give them four to six inches of depth to work with. There should be at least a couple inches between the surface of the soil and the top of the container. Their heavy bodies would suggest they cannot fly, but some of the females can and are attracted to lights. For this reason, there should be a screened cover over the top.

A more interesting cage could be in the form of those used in ant farms. Two rectangular pieces of clear plastic or glass can be connected by the three outer edges. They can either be slid into slots cut into a U-shaped frame or glued/screwed to the outside of that frame. Cover the top with a screened lid of some kind and keep the faces of the container covered with black paper. You can remove the paper when you wish to observe the insect(s) in their burrows.

Food

It is very easy to keep mole crickets well fed. They eat the tender roots of living and dead plants, and a variety of greens, grains, and small creatures. Have a look at what's growing in the vicinity of where you found the crickets, and add some of those plants to the soil. If the plant is too tall for the tank, trim it.

You can add some living creatures such as small earthworms, snails, and beetle grubs. Bear in mind, they will also eat small crickets.

Some fruit left on the surface will be eaten in the evening; strawberries, apples, peaches, and pears are just some examples. Peanuts would be a treat, too. Cornmeal and oat bran sprinkled on the top would be another good addition to the diet.

Keep the soil moist, but don't oversaturate it.

Bush Katydids, Round-headed Katydids, Angle-wings, and True Katydids

The katydids in this group share the same general leaf shape and life histories. Few of the nearly forty species would generally be considered "singers." Most have calls ranging from clicking to croaking. However, many do add a bit of lispy snare to the chorus and can be a joy to listen to! Some of the females call as well, although those calls lack the gusto, and complexity, of the males' calls.

Although the insects in these groups are not trillers, none of them produce what I would consider a grating or unpleasant call. Below I share the song descriptions of some of my favorites. That is not to say that you won't find others to be equally or more gratifying. For me, part of the fun is in getting to know each and every species and the sounds they make.

In addition to their calls, these insects offer a visual aesthetic. They are living, breathing works of nature's art. This should be considered when setting them up in your home. They should be seen as well as heard.

Songs and Range

Northern Bush Katydid *(Scudderia septentrionalus)*—A series of wet ticks followed by a rapid succession of "dzee-dzee-dzee . . ." The ticks sound like a warm-up to the more sustained call. RANGE: northeastern quarter of the United States.

Broad-winged Bush Katydid *(Scudderia pistillata)*—This species has a very complex call. It begins with "zick-zick-zick." It pauses, and then tacks on another, louder, note or two: "zick-zick-zick-ZICK–ZICK." After another pause, more notes, louder yet, are added: "zick-zick-zick-ZICK–ZICK-ZICK–ZICK." RANGE: northern states.

Rattler Round-winged Katydid *(Amblycorypha rotundifolia)*— A high sputtering rattle, often preceded by short sputters. It brings to mind someone trying to start a tiny, flooded, two-cycle engine. RANGE: northeastern quarter of the country.

Greater Angle-wing *(Microcentrum rhombifolium)*—This species has two different calls. One call consists of a series of ticks, sounding like two pebbles being tapped together. The other call is an often repeated, lisping "tzip." RANGE: all parts of the United States except for the Northwest.

Lesser Angle-wing *(Microcentrum retinerve)*—A rapid, lispy rattle. Similar to, but quieter than, the Common True Katydid. RANGE: southeastern states.

Common True Katydid *(Pterophylla camellifolia)*—A loud, croaking "tch-tch-tch . . . tch-tch-tch." Sometimes extra notes will be added, especially in the southern states. RANGE: eastern half of the country, but absent north of Massachusetts.

Home

These insects will spend their time on the leaves and branches within their cage, therefore their domicile will have to be large enough to accommodate twigs with their leaves attached. I prefer to use, at the minimum, the 8-inch by 5-inch plastic terrariums sold at most pet stores. I use larger ones when I wish to include more individuals. The males seem to get along better than field cricket males, and having a couple share a cage has not been a problem.

When cleaning the cage, take care in transferring the katydids into their temporary cage. Many will make a run for it. It's actually more of a jump, or flight. A nearby cat could complicate retrieving the AWOL insect. I usually have a new cage already set up so I only have to move the katydid once.

Food

Katydids eat flowers, leaves, and fruit. Lettuce is always readily accepted, as well as the following leaves: sassafras, cherry, birch, beech, apple, oak, witch hazel, raspberry, dogbane, grape, buckthorn, Joe-Pye-weed, and goldenrod. There are surely many others. It is a safe bet that they will eat the leaf they are found upon. However, it is best to use the younger, more succulent leaves. Clip a branch holding those leaves so it is short enough to lean inside the cage. The leaves and branch should be rinsed and patted dry to remove potential insect predators. Place the cut end of the branch in a florist's stem tube (described in the tree cricket section). The leaves should be replaced when they begin to dry out.

Katydids will eat a wide variety of flowers, including impatiens, goldenrod, daisy, aster, black-eyed susan, and apple. Experiment with different flowers found growing in their area. Fruit always goes over big with katydids. It is their form of dessert, and they will home in on almost any fruit you offer. As with the other insects, I usually put an apple slice on the bottom of the cage. I sprinkle a pinch of crumbled dog or cat kibble on it to add a little protein. Fresh blueberries and strawberries will bring them pure katydid joy! Consider also cucumbers, peaches, pears, and melon.

Provide water with a few spritzes from a water bottle. Spray

Fork-tailed Bush Katydid *(Scudderia furcata)* eating a blueberry.

the leaves, and don't worry about giving the katydids a little mist. In their natural habitat, they get a little wet and it's not bad for them.

Shield-backed Katydids

The shield-backed katydids dwell in the understory of forests and meadows. Despite their somewhat gentle, sputtery call, they can be aggressive predators and will even attack weaker members of their own species. Males are best kept apart in captivity. Care should be taken when handling these insects, as they can bite.

There are over 120 different species in the United States. I have included song descriptions for four of them.

Songs and Range

American Shieldback *(Atlanticus americanus)*—A lazy, but steady, series of high buzzes. RANGE: eastern half of the United States but absent north of Massachusetts.

Protean Shieldback *(Atlanticus testaceus)*—A high-pitched, sputtery buzz. RANGE: eastern half of the United States but absent south of northern Georgia.

Mormon Cricket *(Anabrus simplex)*—A high-pitched, sputtery buzz. RANGE: northwestern quarter of the country.

Roesel's Katydid *(Metrioptera roeselii)*—Either a continuous or broken sputtery buzz. An introduced species. RANGE: northeastern quarter of the United States.

Home

With the exception of Roesel's Katydids, which live in tall grass, these are ground dwellers. I like to provide them with room to crawl on the bottom of the container. They are also insects you want to be able to watch. I love their shapes, and the patterns in Roesel's Katydids make them true works of art! The rectangular aquariums suit these katydids nicely. I use several washed pieces of bark for the ground. They are easy to rinse off when cleaning the cage, and they set off nicely the hardy look of the insects. A few small, carefully selected stones add to the ambiance, as will a little patch of moss. Roesel's Katydids should be kept as described in the following section on meadow katydids.

Food

If you are looking for a place to discard recently swatted mosquitoes, the shield-backeds would be more than happy to help you out. They are predators and hunters of a large variety of insects. They will eat living and recently deceased flies, ants, lacewings, caterpillars, beetles, small grasshoppers, and other insects.

Being omnivores, they also feed on many different herbaceous plants. They prefer the flowers but will also eat the leaves and stems. Strawberries, raspberries, mulberries, blueberries, cherries, and blackberries are among the fruit they feed on. You can also sprinkle some oatmeal or bran flakes on the bottom of the cage.

Give the whole cage a couple of spritzes with a spray bottle every day. Don't be afraid to wet the insect itself.

Meadow Katydids

Ticks, buzzes, and whirs: That's what you can expect to hear from these little sedge mimics. The larger meadow katydids (Orchelimum), having larger stridulating apparatus, will create louder calls. The call of some of the smaller meadow katydids (Conocephalus) can be so soft that it is barely audible. Although both genera share the same requirements, they should not share a cage. The smaller might be eaten by the larger.

There are over forty species in this group, many of them very easy to find in meadows and fields.

Songs and Range

Short-winged Meadow Katydid *(Conocephalus brevipennis)*—One to five soft, liquid "tsicks." RANGE: throughout the eastern half of the country.

Slender Meadow Katydid *(Conocephalus fasciatus)*—A series of soft, locomotive-like ticks. RANGE: entire United States.

Black-sided Meadow Katydid *(Conocephalus nigropleurum)*—A long series of buzzy whirs, blended with rapid ticking. RANGE: the Northeast.

Saltmarsh Meadow Katydid *(Conocephalus spartinae)*—A very soft, sustained, rapid stuttering buzz. This one is my favorite—it's such a gentle call. RANGE: entire eastern and southern U.S. coast.

Woodland Meadow Katydid *(Conocephalus nemoralis)*—Short, gentle buzzes. RANGE: central United States, east to the Atlantic Ocean.

Common Meadow Katydid *(Orchelimum vulgare)*—A sustained buzz that grows louder toward the end. It also ticks. RANGE: much of the eastern half of the country.

Gladiator Meadow Katydid *(Orchelimum gladiator)*—Several "tsicks," followed by a sustained whir that grows louder toward the end. RANGE: northern half of the United States.

Red-headed Meadow Katydid *(Orchelimum erythrocephalum)*—A sharp, wet "tsick," followed by a buzzy burr repeated continually in rapid sequence. RANGE: southeast quarter of the United States, south from New Jersey.

Black-legged Meadow Katydid *(Orchelimum nigripes)*—Quick ticks, followed by a sustained buzzy trill. RANGE: central

United States and south to the border; populations have been
found in the Northeast as well.

Home

Because these katydids feed mostly on grasses and sedges,
the cage should be long enough to lean a handful of blades
and stems inside. A tall jar works well, as does a rectangular
terrarium. I have had males of smaller meadow katydids share a
space, and while there were displays of "back off!" stridulation,
they rarely made physical contact. The orchelimum males are
a bit more aggressive toward one another. I have read they are
cannibalistic, and that the strong will continue to reduce the
numbers of the weak. I have not witnessed this myself, but file
that information away.

Naturally, because the smaller meadow katydids take up less
space, a few more can be added to the mix. I try to keep the
balance so there are always more females than males. The cage
should be large enough to allow them space to keep away from
one another.

Food

The meadow katydids are very easy to keep fed. They eat the
flowers of herbaceous plants and the seeds of grasses, rushes,
and sedges. They also eat the leaves of these plants. Grasses
seem to be favored. The Saltmarsh Meadow Katydids feed on
spartina grass. Just grab a handful of the plants you find them
on, being sure to include the seed or flower heads. Add to that
a couple of different plants from the immediate surrounding
area. It is not necessary to pull the plant out by the roots. They
will eat lettuce, but not with the gusto shown by the other
groups I've mentioned.

These katydids are predacious, but insects make up only a
small part of their diet. If you can include freshly killed, or

live, insects, they may eat them. Mosquitoes, caterpillars, small beetles, flies, ants, aphids, and scale insects are a few examples of creatures they've been known to catch.

Give the whole area a few spritzes with a water bottle once or twice a day, depending on how fast the cage dries out.

Coneheads

I had a Robust Conehead *(Neoconocephalus robustus)* as a pet once. My family made me get rid of it. Many of the coneheads crackle and buzz, but none in my area as loudly as the Robust. Outdoors the buzzing is a welcome sound, much like the call of the Dog-day Cicada. It is a sound of summer. Indoors, it is out of context and less appreciated. But it was fun for a day.

The coneheads do produce some pleasant singers, though. None carry a pitch to challenge the crickets, but pitch isn't always needed to comfort the ear. Think of the lapping of waves or raindrops on the roof.

I also like the look of these insects, with their elegant shapes and painted clown faces. They can be a challenge to catch, but once caught they are easy to keep in captivity. I've selected four species whose call one might consider a positive addition to a room.

Songs and Range
Eastern Swordbearer *(Neoconocephalus ensiger)*—A continuous series of rapid, lisping buzzes. It brings to mind the sound of a fast-moving locomotive. RANGE: northern half of the United States except absent in the West.

Long-beaked Conehead *(Neoconocephalus exiliscanorus)*—A steady series of soft, lisping "ziiits." RANGE: eastern half of the country.

Jimmy listening to singing crickets and katydids in author's studio.

Cattail Conehead *(Bucrates malivolans)*—A steady, questioning "tchir-tchir-tchir?" RANGE: hugging the coast from New Jersey to Texas.

Davis's Conehead (*Belocephalus* davisi)—A quivering whir, given in close sequence. RANGE: far southeastern states.

Home

The cage housing a conehead would be no different from that for the meadow katydids. Most in both subfamilies of insects cling to the stems of grasses and sedges. There can be some cannibalism among them, so avoid putting too many together. If you do wish to add two or three to a cage, it should be large enough to give them room to stay away from one another.

Food

Coneheads are primarily grass-seed feeders. They will also eat the seeds of sedges. They generally ignore the leaves. They will occasionally nibble on a leaf of lettuce, but with little zeal. Collect the plants you find them on, being sure to include those with seed heads. Some of the southern species will also feed on palmetto.

These insects can be fun to observe while they eat. Some hold the seed head with their front tarsi while nibbling on the seeds. It is similar to watching someone eat an ear of corn.

Water is provided with a spritz from a water bottle. Hit the grass and the coneheads themselves.

Notes

1. Music Beckons

1. Thomas J. Walker, "Experimental Demonstration of a Cat Locating Orthopteran Prey by the Prey's Calling Song," *Florida Entomologist* 47, no. 2 (1963): 163–165.

2. Darwin Correspondence Project Database, letter no. 2814, http://www.darwinproject.ac.uk/entry-28141.

3. John D. Spooner, "Pair-Forming Phonotaxic Strategies of Phaneropterinae Katydids," *Journal of Orthoptera Research* 4 (1995): 127–129.

4. Bentley B. Fulton, "Rhythm, Synchronization, and Alteration of Stridulation of Orthoptera," *Journal of the Elisha Mitchell Scientific Society* 50 (1934): 263–267.

5. R. D. Alexander, "Sound Communication in Orthoptera and Cicadidae," *American Institute of Biological Science Publications* (Washington, DC) 7 (1960): 38–92.

6. Michael Greenfield and Igor Roizen, "Katydid Synchronous Calling Is an Evolutionary Stable Outcome of Female Choice," *Nature* 364 (1993): 618–620.

2. Why Listen?

1. Harry Piers, "The Orthoptera of Nova Scotia with Descriptions of the Species and Notes on Their Occurrence and Habits," *Transcripts of the Proceedings of the Nova Scotia Institute of Science* 14, no. 3 (1918): 201–356.

3. The Straight-winged Bearers of Swords

1. John Himmelman, *Discovering Amphibians: Frogs and Salamanders of the Northeast* (Camden, ME: Down East Books, 2006).

2. Stephen J. Gould, *The Panda's Thumb* (New York: W. W. Norton, 1980).

4. The Katydids

1. W. S. Blatchley, *Orthoptera of North-Eastern America* (Indianapolis: Nature Publishing, 1920), 459.

2. Quoted in ibid.

5. The Crickets

1. R. D. Alexander, "The Role of Behavioral Study in Cricket Classification," *Systematic Zoology* 11, no. 2 (1962).

2. Bentley Fulton, "The Genus *Anaxipha* in the United States (Orthoptera: Gryllidae)," *Journal of the Elisha Mitchell Scientific Society* 72 (1956): 222–243.

6. The Mighty Cricket Gladiators

1. Hans A. Hofmann and Klaus Schildberger, "Assessment of Strength and Willingness to Fight during Aggressive Encounters in Crickets," *Animal Behavior* 62, no. 2 (2001): 337–348.

2. Hans A. Hofmann and Paul Stevenson, "Flight Restores Fight in Crickets," *Nature* 403 (2000): 613.

7. "Give a Little Whistle"

1. Eraldo M. Costa Neto, "'Cricket Singing Means Rain': Semiotic Meaning of Insects in the District of Pedra Branca, Bahia State, Northeastern Brazil," *Anais da Academia Brasileira de Ciências* 78, no.1 (2006).

2. Charles Lester Marlatt, *The Principal Household Insects of the United States,* U.S. Department of Agriculture Bulletin no. 4 (1896).

3. Peter Kalm, *Travels into North America* (London: T. Lowndes, 1773).

4. Susan Hubbell, *Broadsides from the Other Orders* (New York: Random House, 1993).

5. Henry Colman, *European Life and Manners* (Boston: Charles C. Little and James Brown, 1850).

6. Stephen J. Simpson, Gregory A. Sword, Patrick D. Lorch, and Iain D. Couzin, "Cannibal Crickets on a Forced March for Protein and Salt," *Proceedings of the National Academy of Sciences* 103, no. 11 (2006): 4152–56.

8. A Blade within a Sea of Grass

1. E. M. Walker, "The Crickets of Ontario," *Canadian Entomologist* 36 (1904).

2. John D. Spooner, "Collection of Male Phaneropterinae Katydids by Imitating the Sounds of the Female," *Journal of the Georgia Entomological Society* 3 (1968): 45–46.

9. The Bug People

1. John Alexander Esquemeling, *The Buccaneers of America* (Amsterdam: Jan ten Hoorn, 1678).

2. W. E. Britton and B. H. Walden, *Guide to the Insects of Connecticut*, Connecticut State Geological and Natural History Survey Bulletin no. 16 (1911).

3. Bentley Fulton, *The Tree Crickets of New York: Life History and Bionomics*, New York Agricultural Experiment Station Technical Bulletin no. 42 (1915), 31.

4. Aristotle, *History of Animals*. Available at http://ebooks.adelaide.edu.au/a/ aristotle/history/. Quotations are from bk. 4, chap. 7.

5. The Linnaean Correspondence, an electronic edition prepared by the Swedish Linnaeus Society, Uppsala, and published by the Centre International d'Étude du XVIIIe Siècle, Ferney-Voltaire. http://linnaeus. c18.net/.

6. R.-A. F. de Réaumur, *Mémoires pour servir à l'histoire des insectes*, 6 vols. (Paris: Académie Royale des Sciences, 1734–1742).

7. A. J. Retzius, *Caroli de Geer genera et species insectorum e generosissimi auctoris scxriptis extraxit, degessit, latinae quoad partem reddidit, et terminologiam insectorum Linneanam addidit*. Lipsiae: Cruse.

8. *Encyclopedia Britannica of Arts, Sciences, and General Literature*, 8th ed. (Edinburgh: Adam and Charles Black, 1855), 9:5.

9. Samuel H. Scudder, "The Note of the Katydid," *Psyche* 1, no. 16 (1875): 93–94.

10. Samuel H. Scudder, "Some American Crickets," *Harper's Magazine*, October 1886, 691–696.

11. Ibid., 696.

12. J. R. Matthews, "History of the Cambridge Entomological Club," *Psyche* 82 (1974): 3–37.

13. W. S. Blatchley, "Days of a Naturalist: An Autobiography of W. S. Blatchley," *BIOS* 12, no. 3 (1941).

14. W. S. Blatchley, *Heteroptera, or True Bugs of Eastern North America, with Especial Reference to the Faunas of Indiana and Florida* (Indianapolis: Nature Publishing, 1926).

15. W. S. Blatchley, *Blatchleyana II* (Indianapolis: Nature Publishing, 1940).

16. Albert P. Morse, "Manual of the Orthoptera of New England, Including the Locust, Grasshoppers, Crickets, and Their Allies," *Proceedings of the Boston Society of Natural History* 35, no. 6 (1920): 197–556.

17. James A. G. Rehn, "Morgan Hebard (1887–1946)," *Entomological News* 59, no. 3 (1948).

18. Morgan Hebard, *The Dermaptera and Orthoptera of Illinois* (Urbana: State of Illinois Division of the Natural History Survey, 1934), 125.

19. Daniel Otte, *The North American Grasshoppers,* vol. 1 (Cambridge, MA: Harvard University Press, 1981); J. A. G. Rehn, *A New Monograph of the Orthoptera of North America (North of Mexico),* vol. 1, Academy of National Science Monograph no. 12, Philadelphia (1961).

Glossary

Aedeagus   Reproductive organ in male insects.

Basal   At or near the base.

Basal antenna segment   The lowermost, often widest, segment of the antennae.

Bivoltine   Two generations in a single year.

Cerci   Plural form of *cercus,* one of the paired projections extending from the tip of the abdomen.

Cone   Elongated area extending from the top of the head, most notably in the Neoconocephalus katydids.

Coniferous   Cone-bearing plants, often called evergreens.

Crepitation   A crackling sound produced by some grasshoppers by rapidly opening their wings.

Dactyl   Digging claw on the front leg of a mole cricket.

Deciduous   Pertaining to trees that shed their leaves.

Dorsal   The upper surface or edge.

Femora   Plural for *femur,* the section of leg between the body and tibia—"thigh."

Genera   Plural for *genus.*

Genus   Part of the two-word naming system (binomial) of an organism. The genus name precedes the species name. Example: In the name *Scudderia furcata,* the genus is *Scudderia.*

Herbaceous   A plant bearing leaves and a stem that die at the end of the growing season.

Hind wings   Also called the inner wings, the pair of wings beneath the tegmina used for flight. Sometimes they are reduced or absent.

Lateral   Pertaining to the side.

Metanotal gland   The gland responsible for producing the liquid nuptial treat for the female tree cricket. It is located beneath the wings of the male tree cricket.

Mnemonic   A memory or learning aid; in the case of insect song, a word used to help describe or remember that sound.

Nymph   Immature stage of an insect undergoing incomplete metamorphosis.

Ovipositor   The elongated structure extending from the female's abdomen used to deposit eggs.

Palpi, palps   Plural form of *palpus* or *palp*. The elongated, paired structures emanating from the lower area of an insect's face. They are mainly sensory organs.

Pronotum   The plate that wraps around the top and sides of the thorax.

Pronotal disc   The dorsal surface of the pronotum.

Spermatophore   A packet containing sperm, produced by the male, which is attached to the female's reproductive opening.

Spermatophylax   A gelatinous packet secreted by the male insect and attached to the female's abdomen for consumption.

Spur   A movable spinelike projection on the hind tibia.

Stridulary area/field/organ   The area on the tegmina that contains the file and scraper for producing sound. It is located at the basal section of the tegmina.

Stridulation   The act of engaging the file and scraper in the stridulary area to produce a sound.

Taxonomist   A person who specializes in the classification of organisms into groups on the basis of their structure, origin, and behavior.

Tarsae   Plural for *tarsus,* they are the segmented part of the leg, or "feet," attached to the tibiae.

Tegmina   Plural for *tegmen*. The forewings, also known as upper or outer wings. They are usually thickened and distinctively veined. In katydids they are mostly opaque, green or brown, and wrap around the abdomen. In crickets they are transparent to translucent, sometimes opaque, and usually rest flat on the abdomen.

Tibiae   Plural for *tibia,* the length of leg between the femur and the tarsus.

Tympanum   The hearing organ, or eardrum, located on the front tibiae of most katydids and crickets.

Univoltine   One generation in a single year.

Online Audio Tracklist

The Online Audio Resource can be accessed at: www.cricket radiobroadcast.com.

The first section consists of the calls of 52 species of crickets and katydids calling in their natural environment between June and October in the eastern United States. The calls blend from one insect, or insects, to the next. The second section breaks out the individual songs for the purpose of identifying the singers. They correspond to the list below.

Spring Field Cricket *(Gryllus veletis)*—Killingworth, CT, yard—June

Spring Field Cricket *(Gryllus veletis)* Courtship Call—Killingworth, CT, yard—June

Carolina Ground Crickets *(Eunemobius carolinus)* in lawn—Killingworth, CT, yard—June

Short-winged Meadow Katydid *(Conocephalus brevipennis)*—Pond Meadow Natural Area fen, Killingworth, CT—June

Fall Field Crickets *(Gryllus pennsylvanicus)* with waves—Misquomicut Beach, RI—July

Fall Field Crickets *(Gryllus pennsylvanicus)* with Eastern Chipmunk and Blue Jay—Platt Nature Center, Killingworth, CT, August

Tinkling Ground Crickets *(Allonemobius tinnulus)* with Fall Field Crickets *(Gryllus pennsylvanicus)* and Blue Jay—Killingworth, CT, yard—August

Allard's Ground Crickets *(Allonemobius allardi)* and train—Wallingford, CT—August

Northern Bush Katydid *(Scudderia septentrionalis)*—Killingworth, CT, yard—August

Fork-tailed Bush Katydid *(Scudderia furcata)* with Carolina Ground Crickets *(Eunemobius carolinus)* in background—Killingworth, CT, meadow—August

Southern Ground Cricket *(Allonemobius socius)*—Someone's lawn, Smyrna, DE—August

Curved-tailed Bush Katydid *(Scudderia curvicauda)*—Platt Nature Center, Killingworth, CT—August

Striped Ground Cricket *(Allonemobius fasciatus)*—Pond Meadow Natural Area fen, Killingworth, CT—August

Striped Ground Cricket *(Allonemobius fasciatus)* courtship call—Pond Meadow Natural Area fen, Killingworth, CT—August

Roesel's Katydid *(Metrioptera roeselii)*—Killingworth, CT, yard—August

Four-spotted Tree Cricket *(Oecanthus quadripunctatus)*—Bombay Hook in Smyrna, DE—August

Say's Trig *(Anaxipha exigua)* in marsh grasses—Hurd State Park, East Hampton, CT—August

Beach Trig in marsh *(Anaxipha literina)*—Tybee Island, GA—August

Fast-calling Tree Cricket *(Oecanthus celerinictus)*—sustained call from roadside grass and broken trill in car—Tybee Island, GA—August

Broad-tipped Conehead *(Neoconocephalus triops)* in a tree—Tybee Island, GA—August

Handsome Meadow Katydid *(Phyllopalpus pulchellus)*—calling twice, with Japanese Bush Cricket *(Velarifictorus micado)* and Narrow-winged Tree Crickets *(Oecanthus niveus)* in background—Skidaway Island State Park, GA—August

Broad-winged Tree Cricket *(Oecanthus latipennis)* in raspberry—Altona Marsh in Jefferson County, WV—August

Eastern Swordbearer *(Neoconocephalus ensiger)* in freshwater marsh—Altona Marsh in Jefferson County, WV—August

Red-headed Meadow Katydid *(Orchelimum pulchellum)* in grass—Okefenokee National Wildlife Refuge, GA—August

Two-spotted Tree Cricket *(Neoxabea bipunctata)*—someone's front yard, Smyrna, DE—August

Common True Katydids *(Pterophylla camellifolia)* in thunderstorm with Tinkling Ground Crickets *(Allonemobius tinnulus)* and Oblong-winged Katydids *(Amblycorypha oblongifolia)* in background—Killingworth, CT, yard—August

Common True Katydids (*Pterophylla camellifolia*—notice the extra notes present in the southern song)—Skyline Drive in Shenandoah National Park, VA—August

Slosson's Scaly Crickets (*Cycloptilum slossoni*) in a tree—Skidaway Island State Park, GA—August

Forest Scaly Cricket (*Cycloptilum trigonipalpum*) in marsh—Ocean County, NJ—August

Davis's Conehead (*Belocephalus davisi*) in a freshwater marsh—Okefenokee Swamp, GA—August

Greater Angle-wing (*Microcentrum rhombifolium*) with Eastern Swordbearer (*Neoconocephalus ensiger*) and Narrow-winged Tree Cricket (*Oecanthus niveus*) in background—Bent of the River Audubon sanctuary, Southbury, CT—August

Lesser Angle-wing (*Microcentrum retinerve*) with Jumping Bush Crickets (*Orocharis saltator*) and Long-beaked Conehead (*Neoconocephalus exiliscanorus*) in background—Ocean County, NJ—August

Nebraska Coneheads (*Neoconocephalus nebrascensis*) in grass along road—Skyline Drive, Shenandoah Mountains, VA—August

Black-horned Tree Cricket (*Oecanthus nigricornis*) in goldenrod—Pond Meadow Natural Area fen, Killingworth, CT—August

Common Meadow Katydid (*Orchelimum vulgare*) in sedges—Pond Meadow Natural Area fen, Killingworth, CT—August

Narrow-winged Tree Crickets (*Oecanthus niveus*)—Cockaponset State Forest, Chester, CT—September

Jumping Tree Crickets (*Orocharis saltator*) on shore of Long Island Sound, Saybrook Point, Old Saybrook, CT—September

Narrow-winged Tree Crickets (*Oecanthus niveus*) and Jumping Bush Crickets (*Orocharis saltator*)—Saybrook Point, Old Saybrook, CT—September

Northern Mole Cricket (*Neocurtilla hexadactylla*) with Wood Thrush—Horse Pond, Madison, CT—September

House Crickets (*Acheta domesticus*—two)—Guppies to Puppies pet store, Old Saybrook, CT—September

Long-beaked Coneheads (*Neoconocephalus exiliscanorus*) with Common True Katydids (*Pterophylla camellifolia*) in background—Pond Meadow Natural Area fen, Killingworth, CT—September

Japanese Burrowing Cricket *(Velarifictorus micado)* under my car—Cedarville State Forest, MD—September

Japanese Burrowing Cricket *(Velarifictorus micado)* courtship call—Maryland—September

Columbian Trigs *(Cyrtoxipha columbiana)* chorusing in the treetops—Savannah, GA, neighborhood—September

Sphagnum Ground Crickets *(Neonemobius palustris)* and chickadees—Beckley Bog in Norfolk, CT—September

Round-tipped Conehead *(Neoconocephalus retusus)* in meadow—Bent of the River Audubon sanctuary, Southbury, CT—September

Rattler Round-winged Katydid *(Amblycorypha rotundifolia)* at side of road—Skyline Drive, Shenandoah Mountains, VA—September

Oblong-winged Katydid *(Amblycorypha oblongifolia)*—Fairgrounds in Delaware—September

Pine Tree Crickets *(Oecanthus pini)* on windy day, with Red-breasted Nuthatch—Wellfleet, Cape Cod, MA—September

Salt Marsh Meadow Katydid *(Orchelimum fidicinium)*—Guilford, CT, salt marsh—September

Seaside Meadow Katydids *(Conocephalus spartinae)*—Guilford, CT, salt marsh—September

Southeastern Field Cricket *(Gryllus rubens)* with Southern Ground Cricket *(Allonemobius socius)* and Narrow-winged Tree Crickets *(Oecanthus niveus)* in background—Fairground field in Delaware—September

Black-legged Meadow Katydid *(Orchelimum nigripes)* on sedge—Hurd State Park, East Hampton, CT—September

Davis's Tree Crickets *(Oecanthus exclamationis)*—Connecticut Water Company property, Killingworth, CT—September

Sand Field Crickets *(Gryllus firmus)*—Saybrook Point, Old Saybrook, CT—September

Robust Conehead *(Neoconocephalus robustus)* with plucking Green Frogs—Altona Marsh in Jefferson County, WV—September

Woodland Meadow Katydid *(Conocephalus nemoralis)* in a meadow—Platt Nature Center, Killingworth, CT—September

Snowy Tree Cricket *(Oecanthus fultoni)*—Wellfleet, Cape Cod, MA—October

Common True Katydid *(Pterophylla camellifolia),* slowing down in the cold . . .—Lake Hammonasset, Madison, CT—late October

Jumping Bush Cricket *(Orocharis saltator),* slowing down in the cold . . .—Killingworth, CT, yard—late October

Carolina Ground Cricket *(Eunemobius carolinus),* the last man standing—Killingworth, CT, yard—November

Bibliography

Alexander, R. D., and D. Otte. 1960. Sound communication in Orthoptera and Cicadidae. *American Institute of Biological Sciences* 7:38–92.

———. 1967. Cannibalism during copulation in the Bush Cricket *Hapithus agitator* (Gryllidae). *Florida Entomologist* 50:79–87.

Aristotle. Ca. 343 BC. *Historia Animalium.* Available online at http://ebooks. adelaide.edu.au/a/aristotle/history/.

Bancroft, Hubert Howe. 1889. *History of Utah.* San Francisco: The History Co.

Bell, Paul C. 1979. Rearing the Black Horned Tree Cricket, *Oecanthus nigricornis* (Orthoptera: Gryllidae). *Canadian Entomologist* 111:709–712.

Beutenmuller, William. 1894. Descriptive catalogue of the Orthoptera found within fifty miles of New York City. *Bulletin of the American Museum of Natural History* 6:253–316.

Bland, Roger G. 2003. *The Orthoptera of Michigan—Biology, Keys, and Descriptions of Grasshoppers, Katydids, and Crickets.* East Lansing: Michigan State University Extension.

Blatchley, W. S. 1920. *Orthoptera of North-Eastern America.* Indianapolis: Nature Publishing.

———. 1930. *Blatchleyana.* Indianapolis: Nature Publishing.

———. 1940. *Blatchleyana II.* Indianapolis: Nature Publishing.

———. 1941. Days of a naturalist: An autobiography of W. S. Blatchley. *BIOS* 12 (3).

Britton, W. E., and B. H. Walden. 1911. *Guide to the Insects of Connecticut.* Connecticut State Geological and Natural History Survey Bulletin no. 16.

Cade, W. 1975. Acoustically orienting parasitoids: Fly phonotaxis to cricket song. *Science* 190:1312–13.

Cantrall, Irving J. 1941. *Compendium of Entomological Methods. Part II: Notes on Collecting and Preserving Orthoptera.* Rochester, NY: Ward's Natural Science Establishment.

Chamberlin, W. J. 1970. *Entomological Literature and Nomenclature.* Westport, CT: Greenwood Press.

De Geer, C. 1773. *Mémoires pour servir a l'histoire de insectes.* 7 vols. Stockholm: Pierre Hesselberg.

Dethier, V. G. 1992. *Crickets and Katydids, Concerts and Solos.* Cambridge, MA: Harvard University Press.

Dickens, Charles. 1845. *The Cricket on the Hearth.* New York: F. M. Lupton.

Dow, Richard. 1937. The scientific work of Albert Pitts Morse. *Psyche* 44:1–11.

Esquemeling, John Alexander. 1678. *The Buccaneers of America*. Amsterdam: Jan ten Hoorn.

Forrest, T. G. 1991. Mate choice in Ground Crickets (Gryllidae: Nemobianae). *Florida Entomologist* 74 (1): 74–80.

Fulton, Bentley B. 1915. *The Tree Crickets of New York: Life History and Bionomics*. New York Agricultural Experiment Station Technical Bulletin no. 42.

———. 1932. North Carolina's singing Orthoptera. *Journal of the Elisha Mitchell Scientific Society* 47:55–69.

———. 1934. Rhythm, synchronization, and alteration of stridulation of Orthoptera. *Journal of the Elisha Mitchell Scientific Society* 50:263–267.

———. 1956. The genus Anaxipha in the United States (Orthoptera: Gryllidae). *Journal of the Elisha Mitchell Scientific Society* 72:222–243.

Galliart, P., and K. C. Shaw. 1994. The relationship of weight and sound level to continuity of male calling in the Katydid, *Amblycorypha parvipennis*. *Journal of Orthoptera Research* 2:43–45.

Gangwere, S. K. 1961. A monograph on food selection in Orthoptera. *Transactions of the American Entomological Society* 87:67–230

Gould, Stephen J. 1980. *The Panda's Thumb*. New York: W. W. Norton.

Greenfield, Michael, and Igor Roizen. 1993. Katydid synchronous calling is an evolutionary stable outcome of female choice. *Nature* 364:618–620.

Gurney, Ashley B. 1963. Harry A. Allard, naturalist: His life and work (1880–1963). *Bulletin of the Torrey Botanical Club* 91 (2): 151–164.

Gwynne, Darryl T. 2001. *Katydids and Bush-Crickets: Reproductive Behavior and Evolution of the Tettigoniidae*. Ithaca, NY: Cornell University Press.

Hebard, Morgan. 1934. *The Dermaptera and Orthoptera of Illinois*. Urbana: State of Illinois Division of the Natural History Survey.

Helfer, Jacques R. 1987. *How to Know the Grasshoppers, Crickets, Cockroaches and Their Allies*. New York: Dover.

Himmelman, John. 2006. *Discovering Amphibians: Frogs and Salamanders of the Northeast*. Camden, ME: Down East Books.

Himmelman, John, and Michael DiGiorgio. 2009. *Guide to Night-Singing Insects of the Northeast*. Mechanicsburg, PA: Stackpole Books.

Hofmann, Hans A., and Klaus Schildberger. 2001. Assessment of strength and willingness to fight during aggressive encounters in crickets. *Animal Behavior* 62 (2): 337–348.

Hofmann, Hans A., and Paul Stevenson. 2000. Flight restores fight in crickets. *Nature* 403:613.

Howard, L. O., and A. Busck. 1936. Andrew Nelson Caudell. *Proceedings of the Entomological Society of Washington* 38 (3).

Hubbell, Susan. 1993. *Broadsides from the Other Orders*. New York: Random House.

Jaeger, Benedict. 1859. *The Life of North American Insects.* New York: Harper.

Laufer, Bethod. 1927. *Insect-Musicians and Cricket Champions of China.* Chicago: Field Museum of Natural History.

Lawrence, K. O. 1982. A linear pitfall trap for Mole Crickets and other soil Arthropods. *Florida Entomologist* 65:376–337.

Linnaeus, Carl. 1735. *Systema Naturae.* Netherlands: Lugduni Batavorum.

———. 1737. *The Linnaean Correspondence.* Electronic edition prepared by the Swedish Linnaeus Society, Uppsala, published by the Centre International d'Étude du XVIIIe Siècle, Ferney-Voltaire. http://linnaeus.c18.net/.

Lü, Bi. 17th cent. *Ming Chao Xiao Shi* (The Minor History of the Ming Dynasty).

Mallis, Arnold. 1971. *American Entomologists.* New Brunswick, NJ: Rutgers University Press.

Marlatt, Charles Lester. 1896. *The Principal Household Insects of the United States.* U.S. Department of Agriculture Bulletin no. 4.

Masaki, S., and T. J. Walker. 1987. Cricket life cycles. *Evolutionary Biology* 21:349–423.

Morse, Albert P. 1920. Manual of the Orthoptera of New England, including the locust, grasshoppers, crickets, and their allies. *Proceedings of the Boston Society of Natural History* 35 (6): 197–556.

Neto, Eraldo M. Costa. 2006. "Cricket singing means rain": Semiotic meaning of insects in the district of Pedra Branca, Bahia State, northeastern Brazil. *Anais da Academia Brasileira de Ciências* 78 (1).

Nickle, D. A. 1984. *Metrioptera roeselii,* new record, a European katydid found for the first time in Pennsylvania (Orthoptera: Tettigoniidae: Decticinae). *Proceedings of the Entomological Society of Washington* 86:744.

Nickle, D. A., and J. L. Castner. 1984. Introduced species of mole crickets in the United States, Puerto Rico, and the Virgin Islands (Orthoptera: Gryllotalpidae). *Annals of the Entomological Society of America* 7:450–465.

Otte, D. 1992. Evolution of cricket songs. *Journal of Orthoptera Research* 1:25–49.

Otte, D., D. C. Eades, and P. Naskrecki. 2001+. Orthoptera Species File Online (Version 2). http://orthoptera.speciesfile.org/HomePage.aspx.

Phillips, Maurice E. 1965. James Abram Garfield Rehn (1881–1965). *Entomological News* 76 (3).

Piers, Harry. 1918. The Orthoptera of Nova Scotia. *Transactions of the Nova Scotia Institute of Science* 14 (3): 201–356.

Rehn, James A. G. 1948. Morgan Hebard (1887–1946). *Entomological News* 59 (3).

Rehn, James A. G., and Morgan Hebard. 1914. Studies in American Tettigoniidae (Orthoptera). *Transactions of the American Entomological Society* 40:271–344.

————. 1914. Studies in American Tettigoniidae: 2. A synopsis of the species of the genus Amblycorypha found in America north of Mexico. *Transactions of the American Entomological Society* 40:315–344.

Scudder, Samuel H. 1875. The note of the Katydid. *Psyche* 1 (16): 93–94.

————. 1886. Some American crickets. *Harper's Magazine,* October, 691–696.

————. 1897. *Guide to the Genera and Classification of North American Orthoptera.* Cambridge, MA: Edward W. Wheeler.

Simpson, Stephen J., Gregory A. Sword, Patrick D. Lorch, and Iain D. Couzin. 2006. Cannibal Crickets on a forced march for protein and salt. *Proceedings of the National Academy of Sciences* 103 (11): 4152–56.

Smith, Ray. Ed. 1973. *History of Entomology.* Palo Alto, CA: Annual Reviews.

Smyth, John. 1784. *A Tour of the United States of America.* London: G. Robinson.

Spooner, John D. 1968. Collection of male Phaneropterinae katydids by imitating the sounds of the female. *Journal of the Georgia Entomological Society* 3:45–46.

————. 1995. Pair-forming phonotaxic strategies of Phaneropterinae katydids. *Journal of Orthoptera Research* 4:127–129.

Vickery, V. R., and D. K. Kevan. 1993. *The Insects and Arachnids of Canada. Part 14: The Grasshoppers, Crickets, and Related Insects of Canada and Adjacent Regions.* Ottawa: Canada Communication Group.

Walker, Thomas J. 1963. Experimental demonstration of a cat locating Orthopteran prey by the prey's calling song. *Florida Entomologist* 47 (2): 163–165.

————. 1969. Acoustic synchrony: Two mechanisms in the Snowy Tree Cricket. *Science* 166: 891–894.

————. 2009. Singing Insects of North America. http://entomology.ifas.ufl.edu/walker/buzz/.

Walker, Thomas J., T. G. Forrest, and J. D. Spooner. 2003. The Rotundifolia complex of the genus Amblycorypha (Orthoptera: Tettigoniidae: Phaneropterinae): Songs reveal new species. *Annals of the Entomological Society of America* 96:443–457.

Weiss, Harry. 1936. *The Pioneer Century of American Entomology.* New Brunswick, NJ: Author.

Wineriter, Susan A., and Thomas J. Walker. 1988. Group and individual rearing of Field Crickets (Orthoptera: Gryllidae). *Entomological News* 99 (1): 53–62.

Zim, Herbert S., and James Gordon Irving. 1951. *Insects: A Guide to Familiar American Insects.* New York: Golden Press.

Acknowledgments

This book came about at the urging of Ann Downer. At the time I had a lot of information on how to identify crickets and katydids, and she set me on the course of delving more deeply into their lives. She and my agent, Russell Galen, were responsible for finding a nice home for this book at Harvard University Press.

Along the way, I spent quite a bit of time exploring the lives of these insects with fellow naturalist Michael DiGiorgio, the artist for *Guide to Night-Singing Insects of the Northeast*. His enthusiasm for these creatures continued to fuel my own.

I spent a good deal of time at museums, most notably the Academy of Natural Sciences in Philadelphia, which not only houses an extensive collection of Orthoptera specimens, but also maintains a well-stocked library. Jason Weintraub and Greg Cowper were nice enough to keep pointing me in the right direction.

Most authors, if allowed, would spend an infinite amount of their time on the research for their books. This author is no exception. The real fun was in seeking out these insects throughout the eastern United States. This often meant acquiring special permission to visit parks and sanctuaries after hours. The following people kept their "doors" open a little later for me: Sara Aicher of the Okefenokee Swamp National Wildlife Refuge, Georgia; Holly Holdsworth of Skidaway Island State Park, Georgia; Robert Cantin Jr. of Cedarville State Forest, Maryland; Amy Pimarolli of The Nature Conservancy, West Virginia; Brian Braudis of the E. B. Forsythe National Wildlife Refuge, New Jersey; Tina Watson and Frank Polyak of the

Bombay Hook National Wildlife Refuge, Delaware; and David Gumbart of The Nature Conservancy, Connecticut, who got us into the Beckley Bog.

Members of Connecticut's Corps of Discovery are always on hand to take on a new mission, and they helped me find some of the critters I sought.

Michael Fisher took on the editing for this book, and his gentle urging brought it to completion, just a *little* late . . .

My wife, Betsy, is so tired of hearing me talk about crickets and katydids now, but has probably taken in so much subliminal information (subliminal, because I'm sure she's stopped listening to me), she could write her own book on the subject. At last, we can sit out on the back deck, relax, and enjoy the concert without me feeling the need to gather material for a book.

Index

Page references in *italics* indicate
photographs and illustrations in the text.